NETWORKING AND ONLINE GAMES

NETWORKING AND ONLINE GAMES

UNDERSTANDING AND ENGINEERING MULTIPLAYER INTERNET GAMES

Grenville Armitage,
Swinburne University of Technology, Australia
Mark Claypool,
Worcester Polytechnic Institute, USA
Philip Branch,
Swinburne University of Technology, Australia

Telephone (+44) 1243 779777

Email (for orders and customer service enquiries): cs-books@wiley.co.uk
Visit our Home Page on www.wiley.com

Other Wiley Editorial Offices

John Wiley & Sons Inc., 111 River Street, Hoboken, NJ 07030, USA

Jossey-Bass, 989 Market Street, San Francisco, CA 94103-1741, USA

Wiley-VCH Verlag GmbH, Boschstr. 12, D-69469 Weinheim, Germany

John Wiley & Sons Australia Ltd, 42 McDougall Street, Milton, Queensland 4064, Australia

John Wiley & Sons (Asia) Pte Ltd, 2 Clementi Loop #02-01, Jin Xing Distripark, Singapore 129809

John Wiley & Sons Canada Ltd, 22 Worcester Road, Etobicoke, Ontario, Canada M9W 1L1

Wiley also publishes its books in a variety of electronic formats. Some content that appears in print may not be available in electronic books.

Library of Congress Cataloging-in-Publication Data:

Armitage, Grenville.
Networking and online games : understanding and engineering multiplayer Internet games / Grenville Armitage, Mark Claypool, Philip Branch.
p. cm.
Includes bibliographical references and index.
ISBN-13: 978-0-470-01857-6 (cloth : alk. paper)
ISBN-10: 0-470-01857-7 (cloth : alk. paper)
1. Computer games – Programming. 2. TCP/IP (Computer network protocol) 3. Internet games. I. Title: Understanding and engineering multiplayer Internet games. II. Claypool, Mark. III. Branch, Philip. IV. Title.
QA76.76.C672A76 2006
794.8′1526 – dc22

2006001044

British Library Cataloguing in Publication Data

A catalogue record for this book is available from the British Library

ISBN-13: 978-0-470-01857-6
ISBN-10: 0-470-01857-7

Typeset in 10/12pt Times by Laserwords Private Limited, Chennai, India
Printed and bound in Great Britain by Antony Rowe Ltd, Chippenham, Wiltshire
This book is printed on acid-free paper responsibly manufactured from sustainable forestry in which at least two trees are planted for each one used for paper production.

Contents

Author Biographies

Grenville Armitage Editor and contributing author Grenville Armitage is Director of the Centre for Advanced Internet Architectures (CAIA) and Associate Professor of Telecommunications Engineering at Swinburne University of Technology, Melbourne, Australia. He received his Bachelor and PhD degrees in Electronic Engineering from the University of Melbourne, Australia in 1988 and 1994 respectively. He was a Senior Scientist in the Internetworking Research Group at Bellcore in New Jersey, USA (1994 to 1997) before moving to the High Speed Networks Research department at Bell Labs Research (Lucent Technologies, NJ, USA). During the 1990s he was involved in various Internet Engineering Task Force (IETF) working groups relating to IP Quality of Service (QoS). While looking for applications that might truly require IP QoS he became interested in multiplayer networked games after moving to Bell Labs Research Silicon Valley (Palo Alto, CA) in late 1999. Having lived in New Jersey and California he is now back in Australia – enjoying close proximity to family, and teaching students that data networking research should be fascinating, disruptive and fun. His parents deserve a lot of credit for helping his love of technology become a rather enjoyable career.

Mark Claypool Contributing author Mark Claypool is an Associate Professor in Computer Science at Worcester Polytechnic Institute in Massachusetts, USA. He is also the Director of the Interactive Media and Game Development major at WPI, a 4-year degree in the principles of interactive applications and computer-based game development. Dr. Claypool earned M.S. and Ph.D. degrees in Computer Science from the University of Minnesota in 1993 and 1997, respectively. His primary research interests include multimedia networking, congestion control, and network games. He and his wife have 2 kids, too many cats and dogs, and a bunch of computers and game consoles. He is into First Person Shooter games and Real-Time Strategy games on PCs, Beat-'em Up games on consoles, and Sports games on hand-helds.

Philip Branch Contributing author Philip Branch is Senior Lecturer in Telecommunications Engineering within the Faculty of Information and Communication Technologies at Swinburne University of Technology. Before joining Swinburne he was a Development Manager with Ericsson AsiaPacific Laboratories and before that, a Research Fellow at Monash University where he conducted research into multimedia over access networks. He was awarded his PhD from Monash University in 2000. He enjoys bushwalking with his young family and playing very old computer games.

Acknowledgements

We would like to acknowledge the permissions of, and give special thanks to, a number of copyright owners for the use of their images in this book.

Figures 2.3, 2.4 and 2.8–Photos reproduced by permission of William Hunter, "The Dot Eaters: Videogame History 101", http://www.thedoteaters.com

Figure 3.10 Warcraft™ provided courtesy of Blizzard Entertainment, Inc.

Figure 3.20 is reproduced by permission of Tiffany Wolf

Figure 3.21 is reproduced by permission of Konami

Figures 3.22 and 3.23 are reproduced by permission of Adrian Cheok

Figures 3.24(a) and (b) are reproduced by permission of Wayne Piekarski

Figures 3.6, 3.9, 3.11, 3.14, and 3.15 are reproduced by permission of Electronic Arts

Figures 2.11, 3.4, 3.5, 7.5, 7.6, and 7.7 are reproduced by permission of Id Software, Inc.

Figure 3.12 Pole Position and Figure 3.13 Ridge Racer™ provided courtesy of Namco®

We would also like to acknowledge the work of Warren Harrop and Lawrence Stewart in constructing a large collection of client-side cheat scenarios from which Figures 7.5, 7.6 and 7.7 were selected.

1

Introduction

A lot has happened since 1958 when William A. Hinginbotham used an oscilloscope to simulate a virtual game of tennis. Computing technology has made staggering leaps forward in power, miniaturisation and sophistication. High speed international data networks are part of modern, everyday life in what we call 'the Internet'. Our peculiarly human desire for entertainment and fun has pushed the fusion and evolution of both computing and networking technologies. Today, computer games are sold to an increasingly significant market whose annual revenues already exceed that of the Hollywood movie industry. Multi-player games are making greater use of the Internet and the driving demand for 'better than dial-up' access services in the consumer space. Yet many networking engineers are unfamiliar with the games that utilise their networks, as game designers are often unsure of how the Internet really behaves.

Regardless of whether you are a network engineer, technical expert, game developer, or student with interests across these fields, this book will be a valuable addition to your library. We bring together knowledge and insights into the ways multi-party/multi-player games utilise the Internet and influence traffic patterns on the Internet. Multi-player games impose loads on Internet Service Providers (ISPs) quite unlike the loads generated by email, web surfing or streaming content. People's demand for realistic interactivity creates somewhat unique demands at the network level for highly reliable and timely exchange of data across the Internet – something the Internet rarely offers because of its origins as a 'best effort' service. Game designers have developed fascinating techniques to maintain a game's illusion of shared experiences even when the underlying network is losing data and generally misbehaving.

For those with a background in data networking, we begin with two chapters by Mark Claypool, 'Early Online and Multi-player Games' and 'Recent Online and Multi-player Games', covering the history of computer games and the various ways in which game-related technology has branched out. From the earliest single-player electronic games, through multi-user dungeons and first-person shooters, to today's emerging augmented-reality games and simulation systems, we have come a long way in 40 years. We cover the definition of multi-player networked games and discuss the meaning of peer-to-peer and client–server communication models in the context of game systems. For those readers with a background in game design and development, our next chapter, 'Basic Internet Architecture', provides a refresher and short introduction to the basics of Internet Protocol

Networking and Online Games: Understanding and Engineering Multiplayer Internet Games
Grenville Armitage, Mark Claypool, Philip Branch

(IP) networking. We review the concept of 'best effort' service, IP addressing and the role of transport protocols such as TCP (Transmission Control Protocol) and User Datagram Protocol (UDP) as they pertain to game developers. When you complete this chapter, you will have an understanding of the differences between routing and forwarding, addresses and domain names. You will learn why Network Address Translation (NAT) exists and how it impacts on network connectivity between game players.

Our next chapter, 'Network Latency, Jitter and Loss', should be of interest to all readers. Here we look in detail at how modern IP networks fail to provide consistent and reliable packet transport service – by losing packets or by taking unpredictable time to transmit packets. We discuss how much of this network behaviour is unavoidable and how much can be controlled with suitable network-level technology and knowledge of game traffic characteristics. This leads naturally to Mark Claypool's next chapter, 'Latency Compensation Techniques', where we look at the various techniques invented by game developers to cope with, and compensate for, the Internet's latency and packet loss characteristics. A fundamental issue faced by multi-player online games is that the latency experienced by each player is rarely equal or constant. And yet, to maintain a fair and realistic immersive experience, games must adapt to, predict and adjust to these varying latencies. We look at client-side techniques such as client prediction and opponent prediction, and server-side techniques such as time warping. Compression of packets over the network is introduced as a means to reduce network-induced latency.

Our next chapter, 'Playability versus network conditions and cheats', takes a different perspective. We look at how two separate issues of network conditions and cheating influence player satisfaction with their game experience. First, we look at the importance of knowing the tolerance your players have of latency for any particular game genre. Such knowledge helps game hosting companies to estimate which area on the planet their satisfied customers will come from (and where to place new servers to cover new markets). We discuss existing research in this area and issues to consider when trying to establish this knowledge yourself. Next we look at communication models, cheats and cheat mitigation. Cheating is prevalent in online games because such games combine competitiveness with a sense of anonymity – and the anonymity leads to a lessened sense of responsibility for one's actions. We look at examples of server-side, client-side and network-based cheating that may be attempted against your game, and discuss techniques of detecting and discouraging cheating.

In 'Broadband Access Networks', Philip Branch takes us through a discussion of the various broadband access technologies likely to influence your game player's experiences in the near future. Access networks are typically the congestion point in a modern ISP service; they come in a variety of technologies allowing fixed and wireless connectivity, and have unique latency and loss characteristics. From a high level, we review the architectures of cable modems, Asymmetrical Digital Subscriber Line (ADSL) links, 802.11 wireless Local Area Networks (LANs), cellular systems and Bluetooth.

We then move in an entirely different direction with the chapter 'Where do players come from and when?'. One of the key questions facing game hosting companies is determining where their market exists, who their players are, and where they reside. This has an impact on the time zones over which your help desk needs to operate and the ebb and flow of game-play traffic in and out of your servers. Taking a very practical direction, we first discuss how you can monitor and measure traffic patterns yourself with freely available open-source operating systems and packet sniffing tools. Then we look at existing research on daily and weekly player usage trends, trends in server-discovery probe traffic that hit your server whether people play or not, and note some techniques for mapping from IP addresses to geographical location.

At the other end of the spectrum is the packet-by-packet patterns hidden in packet size distributions and inter-packet arrival times. In 'Online Game Traffic Patterns', we look at how to measure traffic patterns at millisecond timescales, and show how these patterns come about in First-Person Shooter (FPS) games – the most demanding interactive games available. It is at this level that network operators need to carefully understand the load being put on their network in order to properly configure routers and links for minimal packet loss and jitter. We review how typical FPS packet size distributions are quite different in the client-to-server and server-to-client directions, and how server-to-client packet transmissions are structured as a function of the number of clients. Overall this chapter provides great insight into the burstiness that your network must support if you wish to avoid skewing the latency and jitter experienced by every player.

Then in 'Future Directions', Mark Claypool provides general thoughts on some topics relating to the future of online multi-player games. We particularly focus on the use of wireless technologies, automatic configuration of Quality of Service without player intervention, hybrid client–server architectures, cheaters, augmented reality, massively multi-player games, time-shifting games (where you can start and stop at anytime) and new approaches to server discovery.

Finally, in 'Setting up online FPS game servers', we wrap up this book with a practical introduction to installing and starting your own FPS game servers on free, open-source platforms. In particular, we look at the basics of downloading, installing and starting both Wolfenstein Enemy Territory (a completely free team-play FPS game) and Valve's Half-Life 2 (a commercial FPS). In both cases, we discuss the use of Linux-based dedicated game servers, and provide some thoughts on running them under FreeBSD (both Linux and FreeBSD are free, open-source UNIX-like operating systems available for standard PC hardware).

We hope you will find this book a source of interesting information and new ideas, whether you are a networking engineer interested in games or a game developer interested in gaining a better understanding of your game's interactions with the Internet.

Grenville Armitage (author and editor)

2

Early Online and Multiplayer Games

In this chapter, we cover some of the history of early online and multiplayer games. Like most computer systems and computer applications, online games evolved as the capabilities of hardware changed (and became cheaper) and user expectations from those games grew to demand more from the hardware.

Besides being interesting in their own right, examining early online and multiplayer game history can help us understand the context of modern network games. We will deal with the following:

- Introduce important early multiplayer games that set the tone for the networking multiplayer games that would follow.
- Describe early network games that often had a centralised architecture, suitable for the mainframe era in which they were developed.
- Provide details on turn-based games that were popular before low latency network connections were widespread.
- End with popular network games that made use of widespread Local Area Network (LAN) technology.

2.1 Defining Networked and Multiplayer Games

By its very definition, a network game must involve a network, meaning a digital connection between two or more computers. Multiplayer games are often network games in that the game players are physically separated and the machines, whether PCs or consoles or handhelds, are connected via a network. However, many multiplayer games, especially early ones were not network games. Typically, such multiplayer games would have users take turns playing on the same physical machine. For example, one player would take turns fighting alien ships while the second player watched. Once the first player was destroyed or when he/she completed the level, the second player would have a turn. Scores for each player were kept separately. For simultaneous multiplayer play, either cooperatively or head-to-head, each player would see their avatar on the same screen or the screen would be 'split' into separate regions for each player. For example, a multiplayer sports game

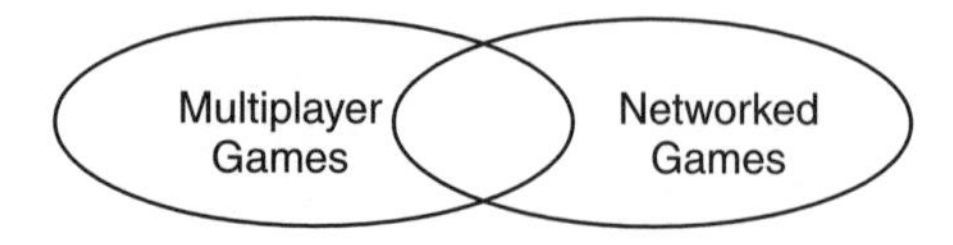

Figure 2.1 The sets of multiplayer games and network games are overlapping, but not subsets or supersets of each other

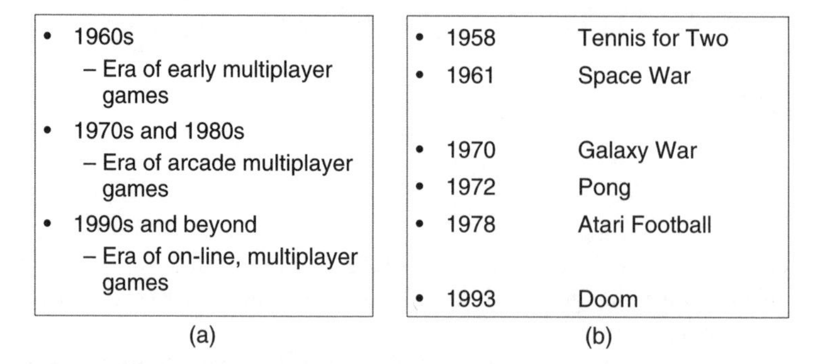

Figure 2.2 Timeline overview of early online and multiplayer games. (a) Lists approximate game eras. (b) Lists the release of milestone games mentioned in this chapter

may have each player working one member of opposing teams. The game field could either be entirely seen by both players or the screen would be physically split into the part of the field viewable by each player. Thus, the area of multiplayer games includes some games that are not network games.

On the flip side, some network games are not multiplayer games. A game can use a network to connect the player's machine to a remote server that controls various gameplay aspects. The game itself, however, can be entirely a single-player game where there is no direct interaction with other players or their avatars. Early games, in particular, were networked because a player logged into a mainframe server and played the game remotely over a network via a terminal. Even with today's modern computer systems, players can run a game locally on a PC and connect to a server for map content or to interact with Artificial Intelligence (AI) units controlled by a server.

Thus, multiplayer and network games overlap, as depicted in Figure 2.1, but neither fully contains the other.

This sets the stage for discussing the evolution of computer games, starting with early multiplayer games, early networked games and progressing to early, multiplayer networked games (Figure 2.2)

2.2 Early Multiplayer Games

In 1958, William A. Higinbotham, working at the Brookhaven National Laboratory, used an oscilloscope to simulate a virtual game of tennis. This crude creation utilised an overhead view, allowing two players to compete against each other in an attempt to sneak the ball past the paddle of their opponent. Higinbotham called this game *Tennis*

Figure 2.3 William Hinginbotham invented the multiplayer game *Tennis for Two* using an oscilloscope. Reproduced by permission of William Hunter.

Figure 2.4 *Spacewar* was the first real computer game, and featured a multiplayer duel of rocket ships. Reproduced by permission of William Hunter.

for Two [PONG] and it was perhaps the first documented multiplayer electronic game (Figure 2.3).

However, while definitely a multiplayer game Tennis for Two used hard-wired circuitry and not a computer for the game play. The honour of the first real computer game goes to *Spacewar*, which was designed in 1961 to demonstrate a new PDP-1 computer that was being installed at MIT (Figure 2.4). In Spacewar, two players duelled with rocket ships, firing torpedos at one another. Spacewar had no sound effects or particle effects, but illustrated just how addictive compelling game play could be even without fancy graphics. It even showed sophisticated AI was not needed since real intelligence, in the form of a human opponent, could enhance game play in both competitive and cooperative modes.

Soon after its creation, Spacewar programmers were discovering the tradeoffs between realism and playability, adding gravity, star maps and hyperspace. Although the price of the PDP-1 (then over $100 000) made it impossible for Spacewar to be a commercial success, it had lasting influence on the games that followed, including subsequent multiplayer and networked games.

A version of Spacewar that was a commercial success was *Galaxy War*, appearing on campuses in Stanford in the early 1970s (Figure 2.5). It may have been up and running even before the far more popular *Pong* by Atari.

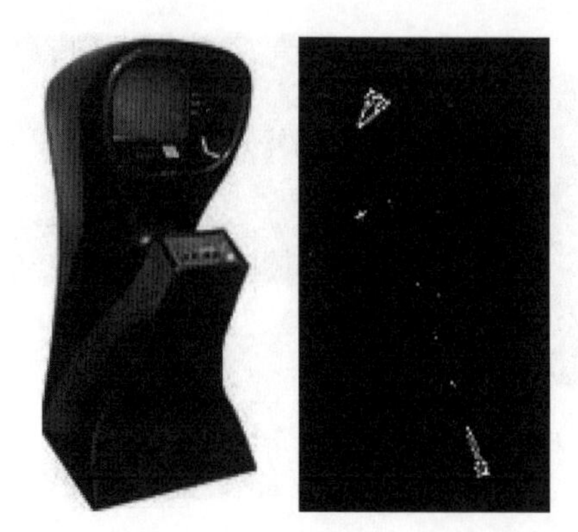

Figure 2.5 *Galaxy War*, early 1970s. Reproduced by permission of Id Software, Inc.

2.2.1 PLATO

Perhaps the first online network community was *PLATO* (which initially was supposedly not an acronym for anything, but later became an acronym for Programming Logic for Automatic Teaching Operations) that had users log into mainframe servers and interact from their terminals [PLATO]. PLATO included various communication mechanisms such as email and split-screen chat and, of course, online games. Two popular PLATO games were *Empire*, a multiship space simulation game and *Airfight*, what may have been the precursor to Microsoft flight simulator. There was even a version of Spacewar written for PLATO. These early online games were networked only in the sense that a terminal was connected to a mainframe, much like other interactive applications (such as a remote login shell or an email client) of the day. Thus, the game architecture featured a 'thin' game client (the terminal) with all the computation and communication between avatars taking place on the server.

The network performance of early systems was thus determined by the terminal communication with the mainframe server via the protocol used by the *Telnet* program [RFC854]. A Telnet connection uses the Transmission Control Protocol (TCP) connection to transmit the data users type with control information. Typically, the Telnet client will send characters entered by keystrokes and wait for the acknowledgment (echo) to display them on the screen. From the user perspective, a typical measure of performance is the *echo delay*, the time it takes for a segment sent by the source to be acknowledged. Having characters echoed across a TCP connection in this manner can sometimes lead to unpredictable response times to user input.

2.2.2 MultiUser Dungeons

MultiUser Dungeons (MUDs) rose to popularity shortly after PLATO, providing a virtual environment for users to interact with the world and with each other with some game-play elements. MUDs are effectively online chat sessions with game-play elements and structure; they have multiple places for players to move to and interact in like an adventure game, and may include elements such as combat and traps, as well as puzzles, spells and even simple economics. Early MUDs had text-based interfaces that allowed players to type in basic commands, such as 'go east' or 'open door' (Figure 2.6). Typically, characters

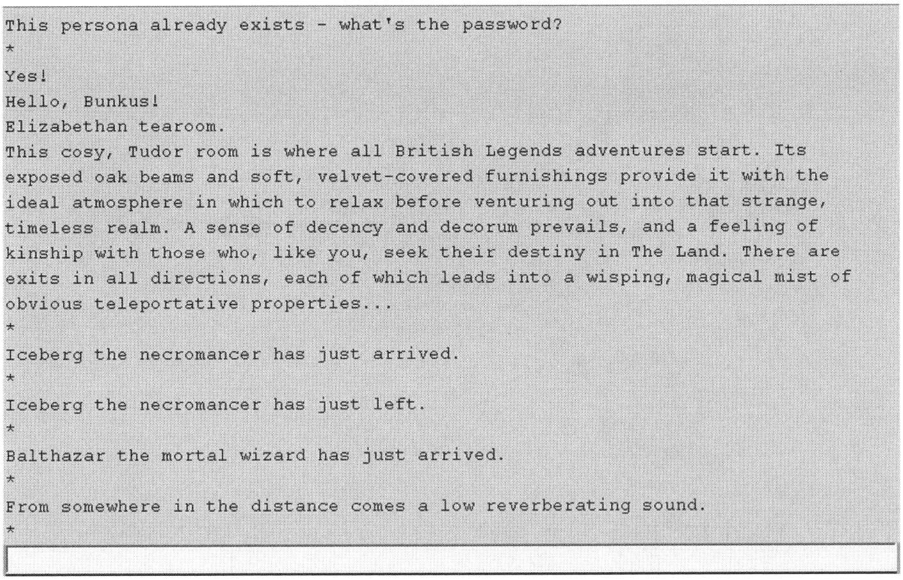

```
This persona already exists - what's the password?
*
Yes!
Hello, Bunkus!
Elizabethan tearoom.
This cosy, Tudor room is where all British Legends adventures start. Its
exposed oak beams and soft, velvet-covered furnishings provide it with the
ideal atmosphere in which to relax before venturing out into that strange,
timeless realm. A sense of decency and decorum prevails, and a feeling of
kinship with those who, like you, seek their destiny in The Land. There are
exits in all directions, each of which leads into a wisping, magical mist of
obvious teleportative properties...
*
Iceberg the necromancer has just arrived.
*
Iceberg the necromancer has just left.
*
Balthazar the mortal wizard has just arrived.
*
From somewhere in the distance comes a low reverberating sound.
*
```

Figure 2.6 Screen shot of MUD 1, one of the early Multiuser Dungeons

can add more structure to the world by adding more content to the world database. The open source nature of many MUDs spurred them on to become popular in academia. Early MUDs became a source of inspiration for later multiplayer network games, such as *Everquest*, and many MUDs still support a core group of dedicated players.

> 'The game was initially populated primarily by students at Essex, but as time wore on and we got more external lines to the DEC-10, outsiders joined in. Soon, the machine was swamped by games-players, but the University authorities were kind enough to allow people to log in from the outside solely to play MUD, as long as they did so between 2 am and 6 am in the morning (or 10 pm to 10 am weekends). Even at those hours, the game was always full to capacity'.
>
> – Richard Bartle, Early MUD history, 15 Nov 90.

MUDs used a client–server architecture, where the MUD administrator would run the server and MUD players would connect to the MUD server with a simple Telnet program, initially from a terminal (Figure 2.7). The disadvantage of Telnet was that it did not always do an effective job of wrapping lines text and incoming messages sometimes got printed in the middle of the commands the user was entering. In response to Telnet's shortcomings, there sprang up a range of specialised MUD client applications that addressed some of the interface issues that Telnet had, and also provide extra capabilities such as highlighting certain kinds of information, providing different fonts and other features.

2.2.3 Arcade Games

Nolan Bushnell, an electrical engineer, was another person influenced by Spacewar, encountering it during the mid-1960s at a university campus in Utah. Apart from just

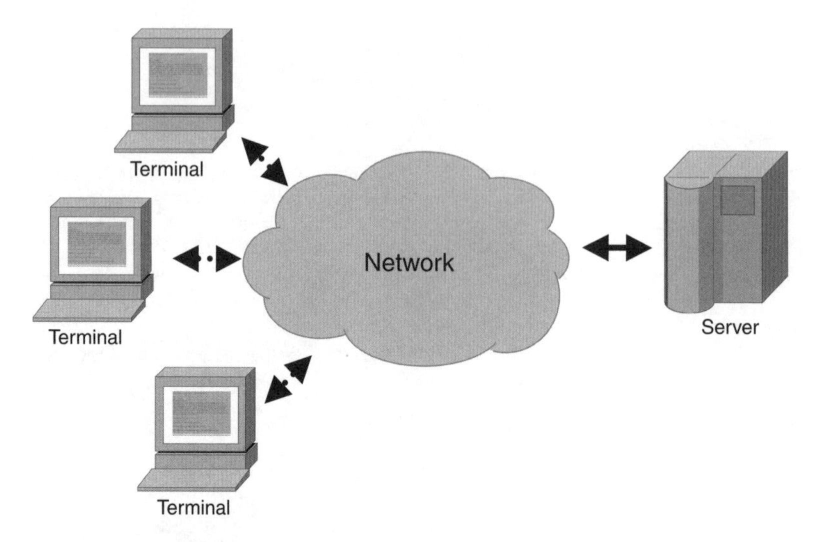

Figure 2.7 Basic client–server topology used by early MUD games

Figure 2.8 *Pong*, one of the best known of the early multiplayer games. Reproduced by permission of William Hunter.

seeing Spacewar as fun, Nolan saw Spacewar as having economic potential. So in 1972, he formed *Atari*, a company dedicated to producing video arcade games. One of the earliest games Atari produced was known as *Pong*, an arcade-friendly version of Hinginbotham's Tennis for Two. Pong was the first big commercially successful video game, while also being a multiplayer game (Figure 2.8).

The 1970s saw a tremendous growth in computer games, with video arcades gaining widespread popularity, and with them, new styles of multiplayer action. In 1978, Atari developed *Football* for video arcades, a game based on American football (not to be confused with European football, also known as *Soccer* in places such as America and Australia), and rendered with simple X's and O's. Atari Football featured addictive multiplayer game play, initially for two players and later for four players (Figure 2.9). Despite the crash in computer games in the early 1980s, Atari released *Guantlet*, an innovative dungeon crawl for up to four players simultaneously (Figure 2.10).

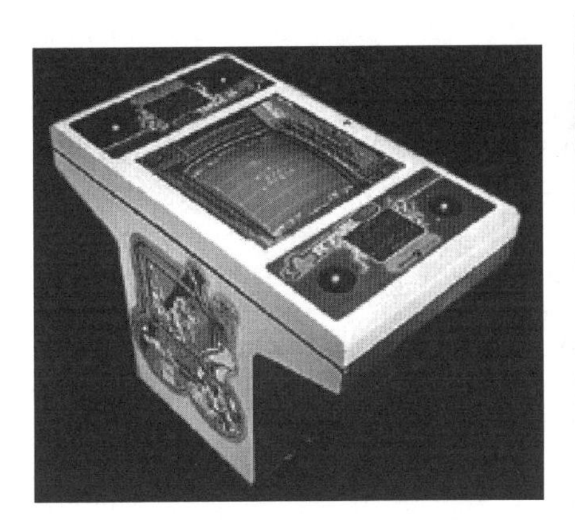

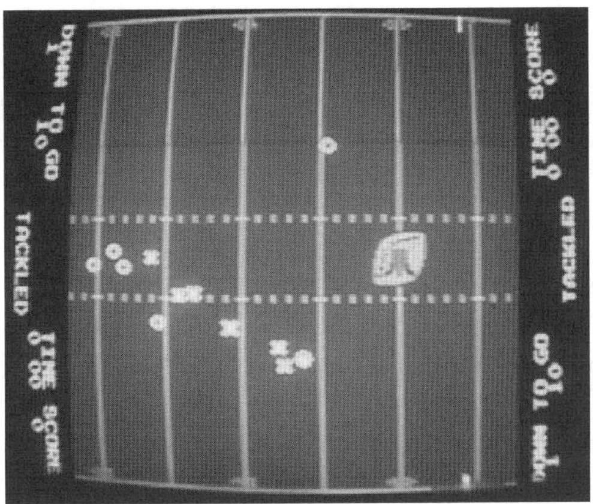

Figure 2.9 Atari *Football* featured two- or four-player multiplayer play

Figure 2.10 *Gauntlet*, released right about the time of the arcade decline, featured two to four players in cooperative, multiplayer hack-and-slash

2.2.4 Hosted Online Games

In the 1980s, the idea of 'pay for play' first emerged, with several game companies hosting online games and charging a monthly fee to play them. Companies such as Dow-Jones (*The Source*) and Compuserv (*H and R Block*) made use of the idle compute-cycles on their servers during nonbusiness hours by charging non-premium fees to access their computers to play games. Such systems primarily featured text-based games that were prevalent in academia, but several were multiplayer versions such as Compuserv's *Mega Wars I*, a space battle that supported up to 100 simultaneous players. Even though such games were limited by today's gaming standards, the prices charged were steep, ranging from \$5 per hour up to \$22.50 per hour.

The high fees for such online game play brought a new group of users who would host individual Bulletin-Board Systems (BBSes) that provided play-by-email or play-by-bulletin-board system versions of table-top games such as chess or Dungeons and Dragons. Users connected to these BBSes by modem (usually by making only a local phone call). Some hobbyists provided richer gaming experiences, still charging money for MUDs, but at much lower, flat monthly rates than had previously been charged.

Although commercial successes, the early computer games were fundamentally different from today's modern computer games. Players would move a dot or simple geometric shape on the screen, perhaps push a button to shoot and something would happen if one shape hit another. There were no opportunities to control anything near human-like avatars, or have complex interactions with other characters or the game-world environment. The game-world environment did not support a variety of vehicles, weapons or even different levels. It is not just that the early games had poorer graphics, rather the game play itself was fundamentally different. Immersiveness, often cited as very important for the success of modern games, was out of the question – a player just controlled abstract shapes on the screen, with any immersiveness coming from the imagination of the player. These early computer games were relatively easy to produce, too, both in terms of cost and time. This is in striking contrast with today's popular computer games, which take 18 to 24 months to produce and often have budgets in millions of dollars.

2.3 Multiplayer Network Games

By the early to mid-nineties, computer power was increasing rapidly, allowing computers to produce more realistic graphics and sound. Computer game players were no longer forced to go to great lengths to suspend their disbelief. Instead of controlling a square moving slowly around on a four-colour screen, they were able to move rapidly in a 256-colour environment, heightening the overall experience of a more realistic, lush, virtual world. In addition, it was increasingly common for computers to have network connections, ushering in a new area in multiplayer games, the multiplayer networked game.

2.3.1 DOOM – Networked First-Person Shooters Arrive

At the end of 1993, id Software produced *Doom*, a First-Person Shooter (FPS) game. Although there had been other FPS games produced before, Doom took the genre to the next level, providing a powerful engine that enabled a fast-paced and violent shoot-'em-up with more realistic levels and creatures than had been seen in previous shooter games (Figure 2.11).

For multiplayer players, Doom enabled up to four players to play cooperatively using the IPX protocol (an early internetworking protocol from Novell) on a LAN, (Figure 2.12) or competitively in a mode that was coined 'death-match'. In the death-match mode, players compete against each other in an attempt to earn more 'frags' (kills) than their opponent(s).

> ***Note:*** Novell's Internet Packet Exchange (IPX), was an internetworking protocol primarily for interconnecting LANs (Figure 2.13). It was often combined with Novell's Sequence Packet Exchange (SPX), to form the SPX/IPX stack – functionally equivalent to the TCP/IP stack on which today's Internet is based. SPX/IPX could not compete with TCP/IP for wide area performance, and has since all but disappeared.

Figure 2.11 Screen shots of *Doom*, the popular First-Person Shooter that started a surge in online, multiplayer gaming. Reproduced by permission of Id Software, Inc.

Software

Doom node		Doom node
IPX driver		IPX driver
Device driver		Device driver
Network card		Network card
Local area network		

Hardware

Figure 2.12 The hardware and software layers required to run multiplayer networked *Doom*

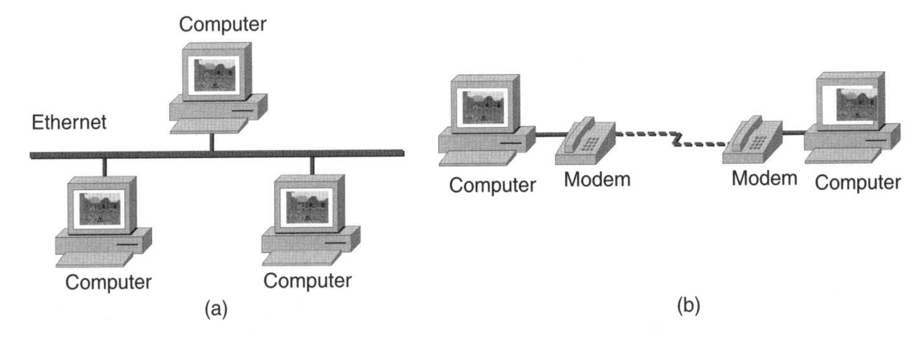

Figure 2.13 Network topologies used by *Doom*. Computers connected to an ethernet LAN acted as 'peers' (a), or computers connected by a modem acted as 'peers' (b)

Doom used a peer-to-peer topology for networking. All players in the game were independent 'peers' running their own copy of the game and communicating directly with the other Doom peers. Every 1/35th of a second, each Doom game sampled the input from each player (such as move left, strafe, shoot, etc.) and transmitted them to all other players in the game. When commands for all other players for that time interval had been received, the game timeline advanced. Doom used sequence numbers to determine if a packet was lost. If a Doom node received a packet number that was not expected (i.e. the previous packet was lost), it decided that a packet had been lost and sent a resend request (a negative acknowledgement, or NACK) to the sender [DOOMENGINE].

Doom peers communicated by using Ethernet broadcasts for all of its traffic. This had the side effect that when a player shot a bullet, the Ethernet packet the Doom peer sent was not only received by all other Doom nodes, but also all other computers on the same LAN were interrupted. The other computers not playing Doom would ignore the broadcast packet, but their processing was still interrupted so they can receive the packet, transfer it to main memory and then have the operating system determine that they do not need it. Normally, LAN traffic is addressed directly to a machine and it is either not received by other machines or it is discarded by the network card before interrupting the processor.

The significant processor time wasted by the computers not participating in a Doom game, but still handling the broadcast packets, was significant, especially for the slower machines of the day, and could even cause them to drop keyboard keystrokes. This was a serious problem for network managers, prompting companies such as Intel and many colleges and universities across the United States to implement specific anti-Doom policies in an attempt to reduce congestion on the local computer networks.

Doom was immensely popular. While the total game sales of 1.5 million copies is not enormous compared with modern blockbuster titles, the shareware version better reflects Doom's popularity. The shareware version of Doom was estimated to have been downloaded and played by 15 to 20 million people [MOD], and installed on more computers than Microsoft's Windows NT and IBMs OS/2 combined. The popularity of multiplayer Doom, particularly the death-match mode, influenced the genre of nearly all FPS games to follow, both in terms of game play and in terms of networking code.

In 1994, id Software produced *Doom 2*, an impressive sequel to Doom. Doom 2 sold over two million copies, making it the highest-selling game by id at that time. Doom 2 could support eight players and, more importantly, Doom's initial use of broadcast packets was removed, and this change brought with it a marked change in the acceptability of networked games on LANs and wide area links.

References

[DOOMENGINE] http://doom.wikicities.com/wiki/Doom_networking_engine, Accessed 2006

[PONG] The Pong Story. The Site of the First Video Game. [Online] http://www.pong-story.com/intro.htm

[MOD] David Kushner. *Masters of Doom*, Random House, 2003. ISBN 1588362892.

[PLATO] Dear, B., PLATO People – A History Book Research Project, http://www.platopeople.com/

[RFC854] Postel, J. and Reynolds, J., "Telnet Protocol Specification", RFC 854. (May 1983)

3

Recent Online and Multiplayer Games

In this chapter, we will deal with the following:

(a) Introduce game communication architectures and their communication models.
(b) Briefly describe the developments in online game play for First Person Shooter (FPS) games, Massively Multiplayer Online games, Real-Time Strategy (RTS) games, and Sports games.
(c) Briefly describe the evolution of game platforms to support online play, including Personal Computers (PCs), Game Consoles and Handheld Game Consoles.
(d) Put games into the broader context of other immersive environments and distributed simulation, including augmented reality (AR), telepresence and virtual reality.

3.1 Communication Architectures

The evolution of online games must consider several different architectures for arranging the communication between game nodes. The different alternatives are depicted in Figure 3.1. The circles represent different processes on remote computers with the links denoting processes that exchange messages.

In the earliest days of multiplayer games, there was no networking between players. Multiplayer functionality was achieved by having both players interacting with the same computer. Players could manipulate their avatars on a shared, common screen or the screen could be physically 'split' by partitioning part of the video screen for each player. Many console games that allow multiple players still use the single screen, multiplayer architecture.

In a *peer-to-peer* architecture, each client process is a peer in that no process has more control over the game than the others. There are no mediator nodes to control game state or route game messages. Peer-to-peer architectures are popular in multiplayer games played on a Local Area Network (LAN) because of the broadcast support of many LANS (e.g. wired or wireless Ethernet) and generally small number of players that participate in a single game. While peer-to-peer architectures can be applied to Wide

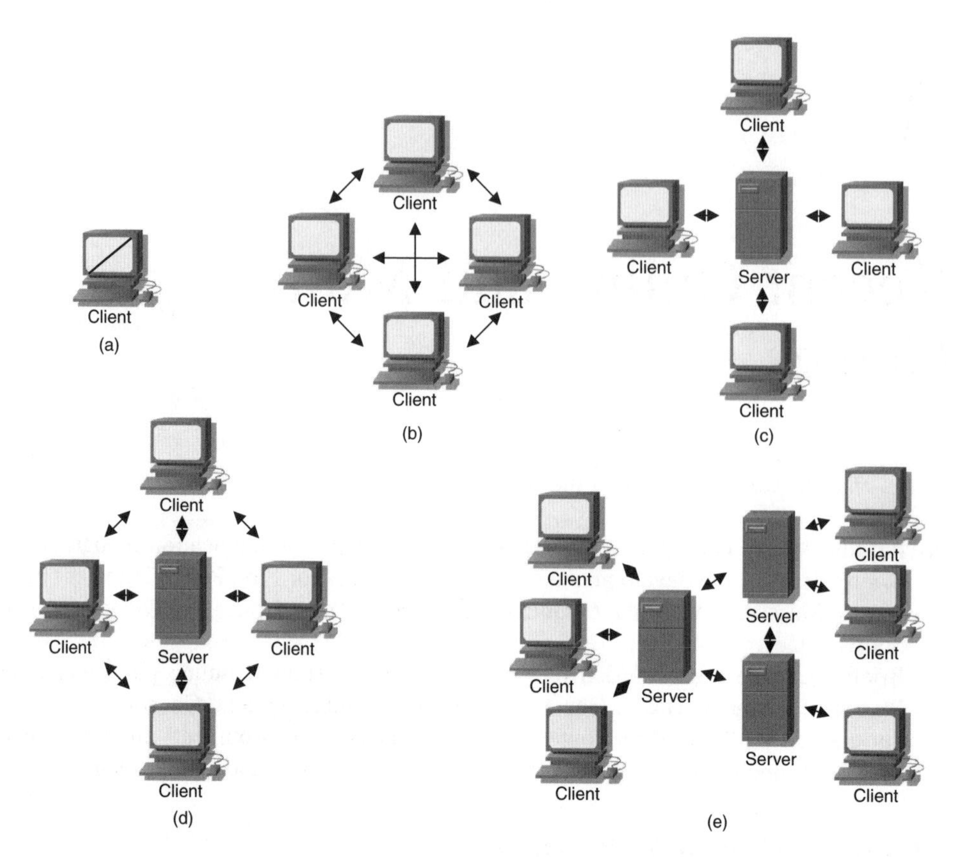

Figure 3.1 The possible communication architectures for multiplayer and network games: a) single screen, possibly 'split', on a single computer, b) peer-to-peer, c) a client–server, d) peer-to-peer-client server hybrid, and e) network of servers

Area Networks (WANs) (i.e. the Internet), they do not scale well without an additional hierarchical structure.

In a *client–server* architecture, one process plays the role of the server, communicating with each client and mediating the game state. The clients do not communicate directly with each other but rather have the server route messages to the appropriate clients. The server is the critical part in the communication link; if a client cannot communicate with the server the game cannot be played and if the server cannot keep up with the communication and computation required, the gameplay can degrade for all clients. The client–server architecture is the most popular architecture used in commercial online games as well as in the classic MultiUser Dungeon (MUD) games (see Chapter 2 for a description of a MUD).

In the *peer-to-peer, client–server hybrid* architecture, the server process mediates game states on the basis of information sent by clients as in the traditional client–server architecture, but the clients are also able to communicate with other clients as in the traditional peer-to-peer architecture. The communication amongst the client peers is generally for game information that is not essential for achieving consistent views of the game state by all clients. For example, it is common to have player Voice over Internet Protocol (VoIP) communication done peer-to-peer with the commands to control an avatar-done client–server.

With a client–server architecture, pure or hybrid, the server can readily become the bottleneck to performance, either because it cannot keep up with the sending and receiving rate for all clients or it cannot process the game-state updates fast enough. With a *network of servers* architecture, the single server can become a pool of several interconnected servers. The communication among the servers can be set up in a peer-to-peer fashion (i.e. all servers are equal) or in a client–server fashion where servers communicate with master servers, obtaining a hierarchical game architecture. By splitting the load from the clients across multiple servers, the network of servers can reduce the capacity requirements imposed on a single server. This can increase the scalability of the game architecture, but has the drawback of a more complicated communication mechanism overall with extra difficulty in keeping game-state information consistent.

When considering communication architectures, it can be useful to differentiate between game system level communication and network level communication. Game system level communication is the manner in which the game elements perceive themselves to be exchanging game-state information and can be both peer–peer and client–server. The network-level communication is how the system-level communication is instantiated when sending data over the Internet, and can also be peer–peer or client–server.

client–server game systems would normally be instantiated by client–server at the network level, but peer–peer at the game system level can be instantiated by peer–peer or client–server at the network level. client–server at the network level is particularly advantageous when the network level server provides minimal processing of game-state information, but otherwise does not parse or modify the game-state messages as it relays between peers. For example, such a server may hide information on Internet Protocol (IP) addresses of the players/clients.

3.2 The Evolution of Online Games

The evolution of online games is best looked at through individual milestones in three of the most popular and influential game genres: *FPS* games, *Massively Multiplayer* games, *RTS* games and *Sports* games (Figure 3.2).

- 1992 Dune II
- 1993 EverQuest
- 1993 Doom
- 1994 Warcraft
- 1994 Doom 2
- 1995 Ultima Online, Asheron's Call
- 1995 Warcraft II
- 1996 Quake
- 1999 Dreamcast (with modem)
- 2000 Xbox and PS2 (with LAN)
- 2005 DS and PSP (with wireless LAN)

Figure 3.2 Timeline overview of online and multiplayer games. The above figure contains notable releases mentioned in this chapter

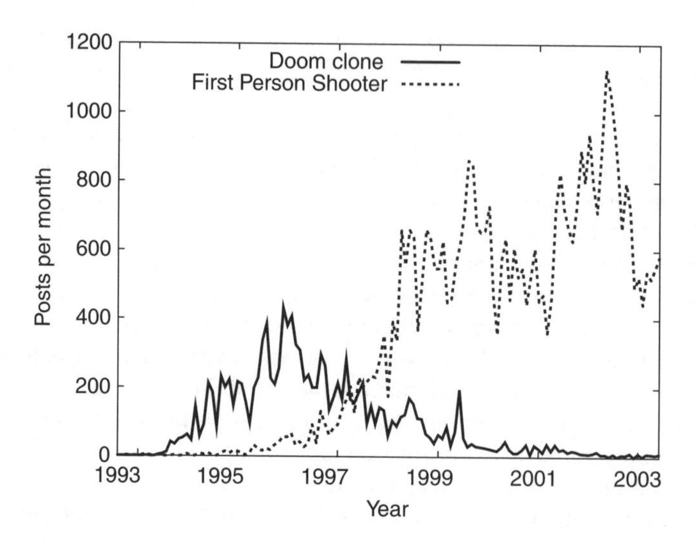

Figure 3.3 Evidence of the popularity and influence of Doom. The term for a game based on Doom (a 'Doom clone') was more common than the now popular term 'First Person Shooter' until the late 1990s

3.2.1 FPS Games

As described in Chapter 2, the advent of Doom brought the genre of the FPS to the forefront as one of the most prominent computer game genres, where it has remained since. Interestingly, the term FPS did not come into popular usage until the late 1990s. The popularity of Doom meant that the phrase 'Doom clones' was more commonly used to refer to what we now called FPS games. Figure 3.3 depicts how pervasive the respective terms were in Usenet news group postings, showing 'Doom clone' was more common than "FPS" until the late 1990s.[1]

By the time of the release of *Doom 2* in 1994, multiplayer network games were generally played by several players over a LAN (see Figure 3.4). For two-player games where the players were not on the same LAN, some games provided support for connections through phone lines by way of a modem or serial cable between the two machines. These early games often used Novell's Internetwork Packet Exchange (IPX) as their networking protocol because of its simplicity. However, IPX was not generally routed on WANs, such as the then-emerging Internet. To overcome this, software such as *Kali* and *iFrag* emerged that allowed IPX to be tunnelled over the Internet (in addition to helping players find other players, a service similar to today's popular *GameSpy*). This tunnelling software enabled players to connect their client PCs to one another for multiplayer network play even though the game software was only designed for LAN play and their PCs were on different LANs. While effective in practice, the early multiplayer network games, such as Doom, were not designed with WAN performance in mind, often suffering when the network capacities were limited or latencies were high.

[1] This data originally appeared at http://en.wikipedia.org/wiki/Doom_clone.

Figure 3.4 Screenshot of Doom 2. Reproduced by permission of Id Software, Inc.

> ***Note:*** Tunnelling is an often-applied technology where a network packet from one kind of network protocol is encapsulated, headers and all, into a data packet in another, lower network protocol. One computer then encapsulates the higher protocol packet, sends (tunnels) it over the lower protocol to a destination where another computer will unpack the higher protocol packet and transmit it normally (see Chapter 4 for more details on IP tunnelling).

Multiplayer network gameplay was significantly improved in 1996 with the release of id Software's *Quake* (Figure 3.5). Quake featured a method that allowed players to compete against each other over the Internet without the need for tunnelling. Before Quake, players needed to coordinate times and places (Internet addresses) to meet online in a game. Quake addressed these problems with the inclusion of servers that stayed-up for repeated rounds, hosting death-match after death-match, so that players from all over the world could connect to these servers at any time of the day or night, and always be able to find a game. The Quake servers acted as persistent game hosts. Players would connect via their Quake clients, with the player's input sent to the server, which would keep track of the state of the game world. Information about the world would be periodically transmitted back to each of the clients, updating their view of the world to match the one currently running on the host machine.

This paradigm created new network problems that had not been faced by game developers before. In particular, the client–server architecture had worked well on a LAN with its high-bandwidth, low-latency connections that were capable of quickly sending and receiving many transmissions for each player. Unfortunately, most WAN (Internet) connections were not capable of LAN transmission speeds, with most players at the time connecting to the Internet through relatively low-speed dial-up modems.

Figure 3.5 Screenshot of Quake. Reproduced by permission of Id Software, Inc.

In 1996, standard-model modems transmitted at 14,400 bps and it took the packets a long time to travel from a client over the modem and across the Internet to a server and back (in games, this time is often referred to as the *ping* time). The time it took to send a command to the server (such as the firing of a gun) and have it result in a change to the game world (such as hitting an opponent) directly affected the realism, immersiveness and overall playability of the game. Players became interested in minimising their ping times. Lower ping times lead to smoother and more immersive gameplay, and, for some games, a higher score (see Chapter 5 for more information on the effects of ping times, i.e. latency, on gameplay).

> ***Note:*** The *ping time* measured by computer games is somewhat different than the ping time measured by network tools such as the network **ping** tool found on many systems. In particular, computer game ping time measures the time to send a User Datagram Protocol (UDP) packet from the client game process to the server game process and back again. Network ping times, on the other hand, measure the time to send an *Internet Control Message Protocol* (ICMP) packet from the client operating system to another host and back again. Game ping times are usually slightly higher than network ping times since they include additional overhead from processing by the server and client applications.

id Software responded by releasing *QuakeWorld* in 1996, a free add-on to Quake that included rewritten network code and a number of game updates [QWORLD]. QuakeWorld was a specific version of Quake optimized for multiplayer Quake play over a modem connection to the Internet. QuakeWorld allowed players to adjust network parameters to minimise the effect of their slow Internet connection on gameplay. In particular,

QuakeWorld implemented a technique known as *client-side prediction*. Clients no longer had to wait for data from the server to update the state of the game world. They were now able to partially predict the future game state, updating it at more regular intervals. Players could then manually set their 'pushlatency', which governed how far in advance many of their clients predicted the game state (see Chapter 6 for more information on latency compensation techniques). QuakeWorld also allowed a player to set a rate limit on the number of packets per second the server would send them, thus avoiding filling up router queues and adding the corresponding latency for a low-bitrate modem connection.

With the release and widespread popularity of QuakeWorld, users all over the world, with different Internet connection speeds were more readily able to play multiplayer Quake with decent performance over the Internet. Organised teams of users called *clans* sprang up, with clans competing other clans, sometimes even in online tournaments for a chance to win cash prizes. Thousands of players competed for the chance to enter id Software's *Red Annihilation* tournament. Multiplayer online gaming even became the full-time job of some gifted players. Dennis 'Thresh' Fong earned well over $100,000 in 1998 competing in Quake tournaments. An organisation known as the *Cyberathlete Professional League* (*CPL*) was started in the late nineties with a goal of bringing in crowds of spectators to watch live death-match tournaments [Kus03].

3.2.2 Massively Multiplayer Games

Besides FPS games, the massively multiplayer on line role-playing (MMORPG) genre started to grow in 1995 with *Ultima Online* (Figure 3.6), a multiplayer network game based on the popular, but single player, *Ultima* series by Origin. Ultima Online initially supported 50 players, a lot at the time but small by today's standards, but was the first

Figure 3.6 Screenshot of Ultima Online. Reproduced by permission of Electronic Arts.

successful game in the MMORPG genre. Microsoft released *Asheron's Call* (with the publishing rights later purchased by the game developer, Turbine) shortly thereafter, along with its online gaming service. At nearly the same time, Sony released *EverQuest*, which soon became the most popular massively multiplayer game thanks in large part to rich graphics and interesting gameplay.

EverQuest had nearly 500,000 subscribers in 1993 [Ken03], and opened the door to dozens of new massively multiplayer online games. Titles such as *Asheron's Call, Dark Age of Camelot* and *Star Wars Galaxies* were follow-on successes to EverQuest. In terms of networking service, these massively multiplayer games charged players a monthly fee (typically around 10 US dollars) to access their characters and the persistent world, which was unique to their commercial successes as opposed to other Internet games where players could play online for free. The Square-Enix's massively multiplayer game *Final Fantasy XI* became the first massively multiplayer game to allow players on PCs and players on game consoles to intermingle in a common world.

An analysis of the growth in massively multiplayer online games, depicted in Figure 3.7 and 3.8, shows a dramatic increase in the number of total subscriptions to MMORPG games since the late 1990s. MMORPG population growth has a hyperbolic or parabolic curve, with little variation in this shape from one MMORPG to another.

3.2.3 RTS Games

The first RTS game released for the computer was Westwood's *Dune II*, released in 1992 [Ger02]. Dune II brought the elements of real-time (as opposed to turn-based) gameplay, with the concepts of building structures with race-specific units and special abilities. Although multiplayer online play was not supported, the potential for multiplayer RTS games had been revealed. Dune II was Westwood's precursor to the popular *Command and Conquer* RTS series (Figure 3.9).

The first RTS game that supported multiplayer online play was Blizzard's 1994 *Warcraft*, that took the Dune II futuristic gameplay to the fantasy world. Although not a big

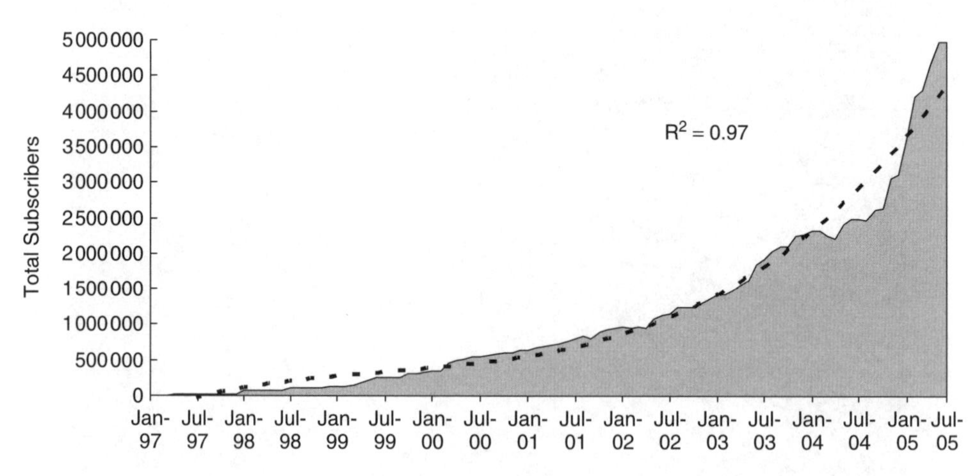

Figure 3.7 Total MMORPG active subscriptions (excluding Lineage, Lineage II, and Ragnarok Online which are much larger) [Woo05]

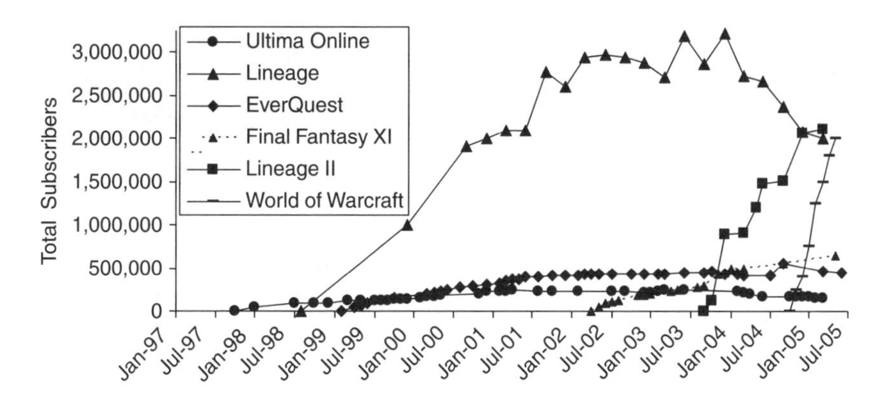

Figure 3.8 Active MMORPG subscriptions for games with 120, 000 + players (excluding City of Heroes, EverQuest II, Dark Age of Camelot, Runescape, Ragnarok Online, Star Wars Galaxies which are all smaller) [Woo05]

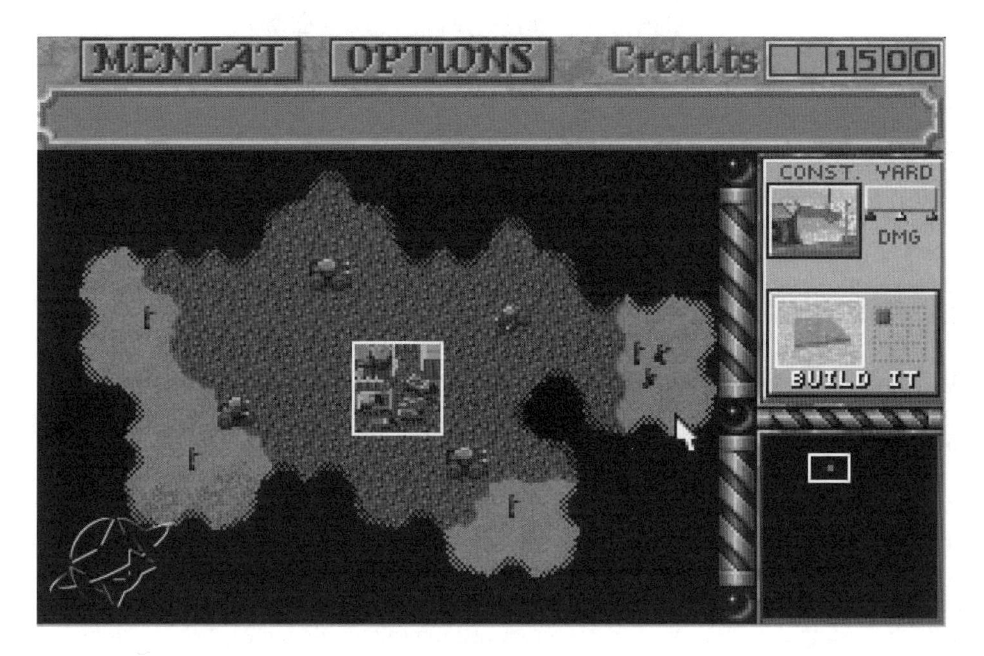

Figure 3.9 Screenshot of Dune II. Reproduced by permission of Electronic Arts.

hit, Warcraft set the stage for Blizzard's *Warcraft II* in 1995, one of the biggest successes the RTS genre has known. WarCraft II allowed up to eight people to play simultaneously on a LAN using the IPX protocol. The v1.2 patch for Warcraft II included optimization for network play, but was not, in fact, playable on the Internet since it still used IPX instead of IP. In the light of this, Blizzard released a special executable to facilitate multiplayer Warcraft II over the Kali network. Warcraft II showed that the Internet could support superb multiplayer RTS gameplay that appeared surprisingly resistant to the effects of latency and bit rate limitations, even on slow modem connections. Because of this, Warcraft II

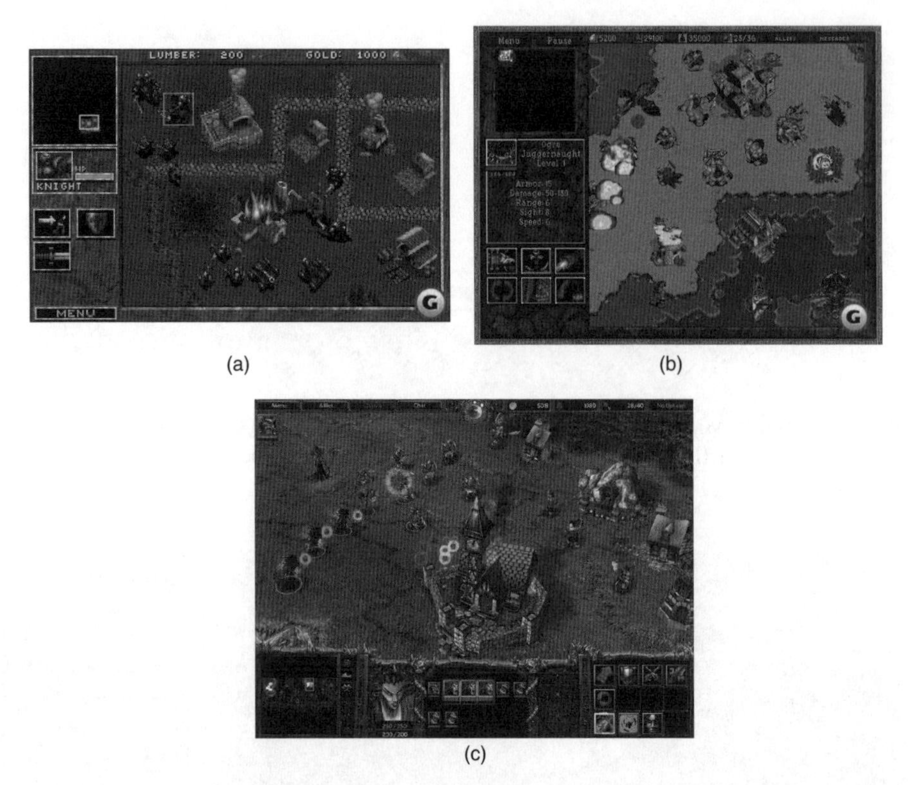

Figure 3.10 Screenshots of Warcraft. Warcraft I is shown in (a), Warcraft II is shown in (b) and Warcraft III is shown in (c). Reproduced by permission of Blizzard Entertainment, Inc.

developed a vigorous online community that even developed into several competitive leagues. Blizzard released Warcraft III in 2002, featuring new gameplay including Heroes and 3-D graphics (Figure 3.10).

The success of multiplayer gaming over the Internet was obvious, so Blizzard introduced *Battle.net*, an online gaming service. Like Kali, Battle.net was a virtual meeting place that permitted players to easily find opponents for Internet play. Use of Battle.net was (and is) free with the purchase of a Blizzard game. Battle.net essentially provided a meeting place for game players, complete with chat rooms and challenge ladders, but without Battle.net actually hosting the game in a client–server fashion (although games such as *Diablo* keep persistent characters and worlds on the Battle.net servers). This kept the bulk of the game traffic from passing through the Battle.net servers, saving on Blizzard's hosting costs and increasing the scalability over a single, centralised client–server model.

3.2.4 Sports Games

Nearly as long as there have been computer games there have been computer games based on sports (see Atari Football in Chapter 2). Sports lend themselves to competition that naturally suggests multiplayer sports games. While the field of sports is nearly as varied as computer games themselves, two multiplayer online games that are popular and have been studied academically are American Football and Car Racing.

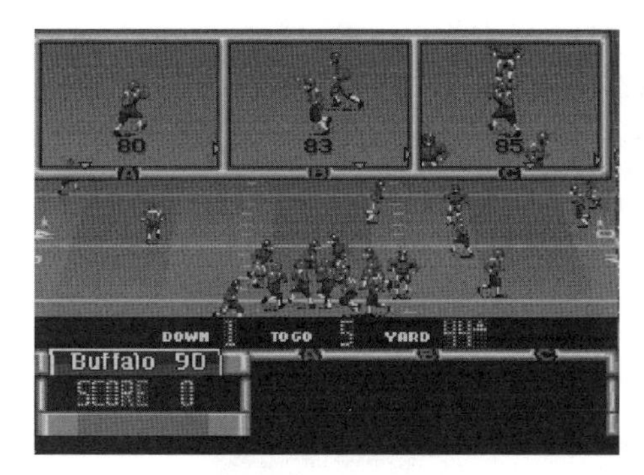

Figure 3.11 Screenshots of Madden NFL. The 1993 Sega Genesis version is shown on the left, the 2005 Microsoft Xbox version is shown on the right. Reproduced by permission of Electronic Arts.

Figure 3.12 Screenshot of pole position. Copyright Namco®.

Establishing itself in the mid-1990s, *Madden NFL Football* (Figure 3.11), in its various versions, is the highest revenue-generating video game franchise in North America and in computer game history [MADDEN05]. Online play, however, was only introduced to the 2003 version and was only available for the Sony Playstation console or a Microsoft Windows PC. As of July 2004, Madden games are also enabled for online play on the Microsoft Xbox Live network. As of 2005, Madden only supports two-player games. Online services (such as for the Playstation or Xbox) via a centralised server enable opponents to locate each other, but players communicate in a peer-to-peer fashion directly with each other [NC04].

The first computer racing game that was released was *Pole Position* (Figure 3.12), popular in the early 1980s because of the quality of the graphics at the time [http://en.wikipedia.org/wiki/Sim_racing]. Although early versions of computer racing games were single player, by the early 1990s, popular series such as *Ridge Racer* and *Need for Speed* had multiplayer support (Figures 3.13 and 3.14). In 2001, Electronic Arts released *Motor City Online* depicted in Figure 3.15, the first online racing game that offered persistent profiles for players. Players earned points and money in racing through different tracks, head-to-head or solo, and could then level up the avatar, and purchase and upgrade vehicles.

Other notable online racing games include the *POD* series that allowed simultaneous play for eight players over the Internet with *POD, Speedzone* being the first online racing

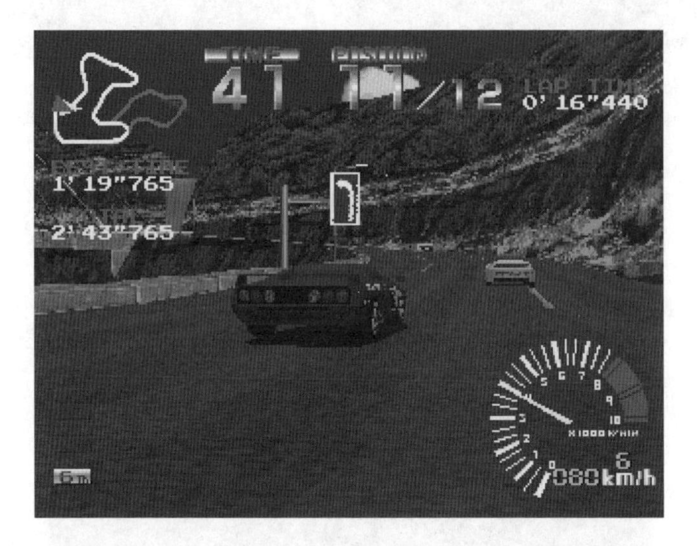

Figure 3.13 Screenshot of Ridge Racer™ (Sony Playstation version). Copyright Namco®.

Figure 3.14 Screenshot of Need for Speed Underground 2. Reproduced by permission of Electronic Arts.

Figure 3.15 Screenshot of Motor City Online. Reproduced by permission of Electronic Arts.

game for game consoles. The game 4 × 4 *Evolution* for the Sega Dreamcast console was the first game to allow online play between Dreamcast, Macintosh and PC platforms.

3.3 Summary of Growth of Online Games

The different game genres can be summarised in the form of a table with some online characteristics:

Genre	Notable release	Max players	Architecture
Real-Time Strategy	Warcraft	About 10	Client–Server
Sports Games	Madden NFL	About 10	Peer-to-Peer
First Person Shooter	Doom	About 50	Client–Server
Online Role Playing	EverQuest	About 10,000	Client–Server

The introduction, growth in popularity and in some cases the decline for each of the genres can be depicted by examining the number of Usenet news posts related to each genre. The data from Figure 3.3 can be combined to have Usenet newsgroup posts on 'Doom clones' and 'FPSs' represent all FPS games. Similarly, the number of Usenet posts[2] related 'MMORPG', 'Real-Time Strategy' and 'Racing Game' can represent the pervasiveness of the Massively Multiplayer, RTS and Sports genres, respectively. Notice that all genres have seen a rise and then subsequent fall in popularity (at least according to Usenet posts in Figure 3.16) with the exception of the MMORGP genre which is still going strong.

[2] Obtained from http://groups.google.com/

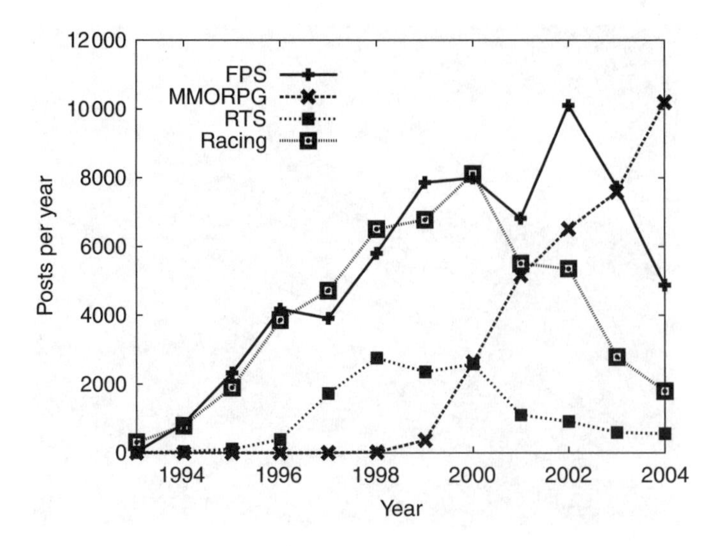

Figure 3.16 The growth in Internet games, depicted by the frequency of occurrence in Usenet news group postings

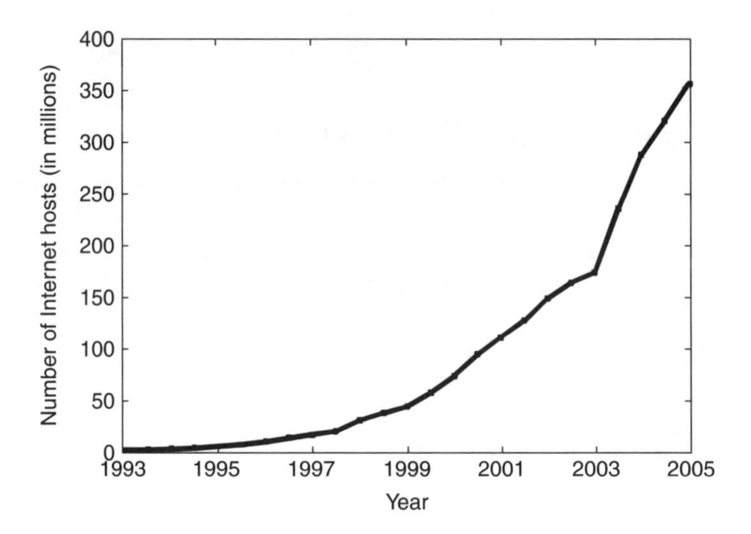

Figure 3.17 The growth of the Internet, depicted by the number of Internet addresses that have been assigned a name

While the growth in the online game genres have been substantial and perhaps even impressive, it is interesting to put the online game growth in the context of Internet growth. Figure 3.17 depicts the growth of the Internet in terms of the number of hosts.[3] The Internet has shown exponential increase in the number of hosts during the same time period, although there are signs that this exponential growth may be slowing.

[3] Data obtained from Network Wizards, http://www.nw.com/

3.4 The Evolution of Online Game Platforms

3.4.1 PCs

The PC has continued to evolve as a general computing platform, as well as a gaming platform, at a phenomenal pace. Improvements to processing power have continued to follow Moore's law, doubling approximately every 18 months. Graphics cards, the core component of any PC used for serious gaming, have done even better, doubling in speed about every 6 months. Random access memory (RAM) speeds and capacities have kept pace with processor improvements. Disk drives that are not generally used during real-time gameplay because their access times are slow, have kept up with the demand for storage capacity increases for games that require more space. Displays have gone from small, monochrome cathode-ray tubes to 24-bit colour, high-resolution, wide-screen liquid crystal display (LCD) displays. Audio has gone from tiny-sounding, integrated PC speakers to surround-sound, 6.1 channel audio. Input devices for PC games are still primarily via a keyboard and mouse, but PCs support game controllers as well as force feedback joysticks (particularly useful for flight simulators).

Figure 3.18 summarises the performance evolution of the personal computer. A 1981 point of reference is provided as a standard[4] computer that had a unit performance of 1. Depicted are power, memory capacity and network capacity and a typical price of about $2500 for a machine.

3.4.2 Game Consoles

By the late 1990s, online gaming really only existed for PCs, with game console systems still being off-line (but certainly multiplayer via split-screen or joint-screen technologies). That all changed as the year 2000 approached. In 1999, Sega introduced the *Dreamcast* that was the first console to include a built-in 56k modem. While the Dreamcast had numerous technically advanced hardware features and even the support of several popular network games such as Quake 3 and Phantasy Star online through the SegaNet gaming service, it was unable to unseat Playstation and Nintendo as the dominant home consoles. The years right after 2000 saw each of the three major consoles (Sony's *Playstation 2*, Microsoft's *Xbox*, and Nintendo's *Gamecube*) equipped with online capabilities (although Nintendo's Gamecube did not feature built-in networking, users were able to buy network adapters that connected via the Gamecube's serial port). By 2004, Microsoft's online live service reportedly had over 1 million subscribers [Tut04].

	1981	2005	Factor
Power	1	1600	1600
$/Power	$100K	$1	100,000
Memory	128K	2G	15,000
Disk capacity	10M	10G	1000
Net bandwidth	9600b/s	1Gb/s	100,000

Figure 3.18 Evolution of the PC computer hardware, from 1981 to 2005

[4] Measured by SPEC, the Standards Performance Evaluation Corporation.

As of 2005, the hardware components of today's consoles are as follows [Tys06]:

Sony PlayStation 2

- Processor: 128-bit "Emotion Engine", 300 MHz
- Graphics: 150 MHz, 4 MB VRAM cache, 75 million polygons per second
- RAM: 32 MB RDRAM
- Other features:
  - Two memory card slots
  - Optical digital output
  - Two USB ports
  - FireWire port
  - Support for audio CDs and DVD-Video

Nintendo GameCube

- Processor: "Gekko" IBM Power PC microprocessor, 485 MHz
- Graphics: ATI 162 MHz, 4 MB RAM, 12 million polygons per second
- RAM: 40 MB (24 MB 1T-SRAM, 16 MB of 100-MHz DRAM)
- Other features:
  - Two flash memory slots
  - High-speed parallel port
  - Two high-speed serial ports
  - Analog and digital audio-video outputs

Microsoft Xbox

- Processor: Modified Intel Pentium III, 733 MHz
- Graphics: nVidia, 250 MHz, 125 million polygons per second
- RAM: 64 MB (unified for audio, video, graphics)
- Network 10/100-Mbps Ethernet, broadband enabled, 56K modem (optional)
- Other features:
  - 8-GB built-in hard drive
  - 5X DVD drive with movie playback
  - 8-MB removable memory card
  - Expansion port

The next generation of consoles promises to be even more powerful [CNET05].

3.4.3 Handheld Game Consoles

Early handhelds were primarily single player. Multiplayer mode was generally enabled by players taking turns, as in Mattel's *Electronic Football*, but some handhelds allowed two players to play simultaneously on the same device, sharing the display. Nintendo's *Game Boy*, released in 1989 and the most popular handheld console ever, was the first handheld console to support networked multiplayer gaming. This was accomplished by connecting a (short) wire between up to four Game Boys, and the game software could even be downloaded from one handheld to another.

The newest generation of handheld gaming systems are equipped with a wireless mode for multiplayer network gameplay. Sony's *Playstation Portable* (PSP), and Nintendo's *Dual-Screen* (DS) support the IEEE 802.11b wireless local area network (WLAN) standard. The Nintendo DS provides a specific subset of the features of 802.11b in order to save battery power for wireless network play. Notably, only a short preamble is used, and only the WLAN capacities of 1 Mbps and 2 Mbps are used, even though 802.11b supports capacities up to 11 Mbps, because these lower capacities consume less power. The Nintendo DS does not support any network layers above the 802.11 layer (in other words, no Internet packets), while the Sony PSP allows and supports Internet packets (IPv4) in the 802.11 infrastructure mode. This allows the PSP to connect to an Access Point (AP) and provide gameplay over the Internet.

The newest release of the PSP even includes a Web browser.

Nintendo DS

System

- Resolution 256 × 192, 260k colors
- Dual ARM processors
- ARM 9 at 67 MHz, ARM 7 at 33 MHz
- 4 MB RAM
- Touch-pad

Network

- WLAN is IEEE 802.11b
- Short preamble
- Wireless capacities of 1 Mbps or 2 Mbps, only
- Ad hoc mode (peer-to-peer)
- No IP stack (UDP/TCP)

Sony PSP

System

- Resolution 480 × 272 pixels, 24-bit color
- Dual MIPS R4000 processors at 333 MHz
- 32 MB RAM
- Analog joystick

Network

- WLAN is IEEE 802.11b
- 1, 2, 5.5, 11 Mbps
- Ad hoc mode (peer-to-peer) and infrastructure mode (Internet)
- IP stack

3.4.4 Summary

The evolution of gaming platforms can be summarised by a table, taking a representative example from each genre and comparing the gaming capabilities:

Platform	Processor	Memory	Resolution	Network
Handheld	333 Mhz	32 MB	480 × 272	WLAN
Console	733 Mhz	512 MB	640 × 480	10/100 Ethernet
PC	3500 Mhz	2048 MB	1600 × 1200	100/1000 Ethernet

3.5 Context of Computer Games

Shared space technologies, including computer games, can be classified according to their *transportation*, the extent to which participants leave behind their local space, and their *artificiality*, the extent to which the space is generated by a computer [BGR+98]. From these two dimensions, broadly, there emerge four classifications of technology (depicted in Figure 3.19):

(a) In *physical reality*, participants reside in the immediate, physical world in that physical game elements are tangible and the participants are corporeal.
(b) With *telepresence*, the participants are transported to a real-world location that is separated from their physical location.
(c) For augmented reality, the local, physical environment is present, but synthetic, computer-generated objects are overlaid on the local environment.
(d) *Virtual reality* allows the participants to be immersed in a remote, synthetic world. Computer games typically are completely artificial in that the game world is entirely generated by a computer and completely transportive in that the player controls an avatar immersed in the game world.

3.5.1 Physical Reality

Physical reality encompasses interactions with objects in the world that take place in everyday life. Games based on physical reality have the participants in the same location,

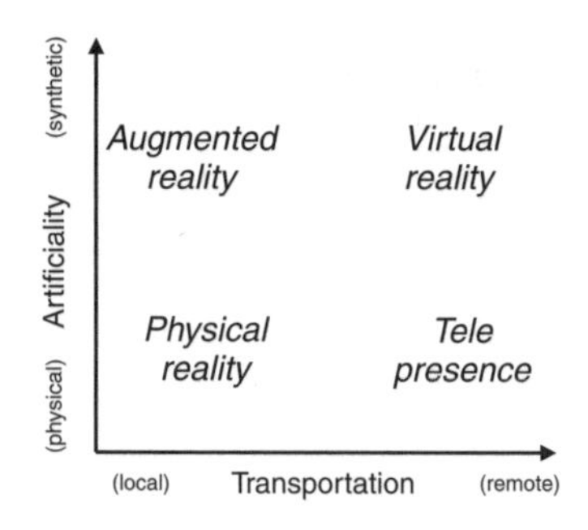

Figure 3.19 Classification of shared spaces according to transportation and artificiality (from [BGR+98])

Figure 3.20 Real people playing Rugby. Do not underestimate the immersiveness of Physical Reality!. Reproduced by permission of Tiffany Wolf.

with the game objects being tangible and concrete. The variety of such physical reality games range from sports, such as cricket and rugby (Figure 3.20), to board games such as chess and backgammon. While games in this cluster are generally not computer supported, there can be computers enhancing aspects of the game. For example, sensors and computer equipment allow accurate calling of the boundary lines in professional tennis. In addition, asynchronous communication by computers can allow games to be played without participants being in the same physical location. For example, chess can be played by two people who are geographically separated and use email or even VoIP to communicate their moves to their opponent.

3.5.2 Telepresence

Telepresence uses computers and networks to provide a feeling of immersiveness for a user that is physically separated from the environment. Sensors gathering information on the local environment transmit information to a physically separated user, providing feedback that allows some recreation of the local world from the remote location.

With telepresence, managing latency is critical for providing a feeling of immersiveness. In the local world, an action has an immediate consequence, whereas in telepresence, an action done remotely is delayed by the time it takes to travel from the remote user to the local environment, plus any additional processing time. Force feedback, where tension or pressure occurs from physically manipulating the local environment, is also often necessary to make the remote user immersed in the local world and can often help overcome deficiencies in the display from a low resolution or frame rate [MS94].

Examples of telepresence include undersea work to repair pipelines and cabling system. Deep water diving is risky and costly, so telepresence systems can prove cheaper as well as posing fewer risks to humans. Similarly, other hazardous environments such as explosive disposal, rescue work and even mining can benefit from telepresence systems.

Another example of telepresence is computer-assisted surgery, which is gaining in popularity because it allows specialists to provide medical care to a geographically broader audience, as well as providing opportunities for computers to make surgeons more precise. When there are concerns about hazardous environments, this brings telepresence to the front line of battlefields, allowing surgeons to aid soldiers during combat.

Telepresence in education can allow people to experience environmentally fragile locations, such as the coral reefs or Egyptian tombs. Space exploration can also benefit from telepresence, as in operating a rover on Mars from Earth [http://marsrovers.jpl.nasa.gov/home/]. Telepresence systems could even be incorporated in entertainment systems beyond games, by allowing the experience of a roller coaster or sky-dive without the associated risks.

Telepresence has even been applied to human-animal communication. A system has been developed that allows humans to remotely 'touch' poultry by having the animal wear a lightweight, tactile pet jacket. The human interacts with a physical representation of the animal (a chicken-robot) and the feedback from the robot is passed through the Internet to the real chicken [PCS+06].

3.5.3 Augmented Reality

With Augmented Reality (AR), the users can still interact with the local, physical environment but artificial, computer-generated objects and attributes are used to enhance the local reality.

A popular game that is a step toward AR is the arcade hit *Dance, Dance Revolution* (*DDR*), initially released by Konami in 1998 with numerous variations available in the arcade, home PCs and game consoles. In DDR, players move their feet to a set pattern, stepping on arrows on the floor to the rhythm of the song. Multiplayer versions allow two players to play cooperatively or in competition. In some sense, DDR as in Figure 3.21 provides a novel interface to a computer game, but in another sense, it augments the reality of dancing to music by providing visual feedback on the dancing itself as gameplay.

As an example that augments reality even more, consider the original *Pacman* game that featured the well-known, cookie-shaped disk that game players manoeuvred through maze munching dots. An AR version of the game called *Human Pacman* depicted in Figure 3.22 [CFG+03] has game players assume the role of the gobbling disk where the physical world is augmented with virtual dots the players see by means of a wearable computer with a head-mounted display. Sensors on the display indicate the direction the player is looking and the GPS provides positional information to generate virtual dots at the appropriate physical locations. Multiplayer aspects are incorporated by having some players assume the role of the ghosts that chase Pacman. Bluetooth technology provides networking to enable the Pacman player to 'pickup' power pills to then chase the ghosts. Human Pacman goes beyond basic Pacman gameplay of one yellow disc against all the ghosts by enabling an additional player to see an overview of the game field on a networked computer, then relaying information about the location of ghosts and power pills to the human Pacman (Figure 3.23).

FPSs have also entered the AR game space, notably with *ARQuake* an AR version of the popular Quake game [PT02]. ARQuake players as in Figure 3.24 use a head-mounted

Figure 3.21 Picture of an arcade version of Dance, Dance Revolution. Reproduced by permission of Konami.

display, mobile computer, head tracker, and GPS system to provide inputs to the game computer, allowing the player to walk around the real, physical world and fight against virtual Quake monsters.

A game called *Real Tournament*[5] is also an AR shooting game. However, instead of a head-mounted display, Real Tournament (Figure 3.25) attaches an IPAQ Pocket PC with a 802.11 WLAN network card to a super-soaker squirt gun. A microcontroller resolves communication between the trigger, the GPS and an electronic compass and sends the information to the IPAQ. The IPAQ also provides team communication by a push-to-talk audio feature over the wireless network, allowing players to synchronise their actions. The network includes a wireless network around parts of a University campus, based on IEEE 802.11b and GPRS. The IPv6-enabled IPAQs are mobile, allowing players to transparently roam between the wireless networks without losing connectivity or disrupting the game state.

[5] The name is a play of words from Epic's Unreal Tournament.

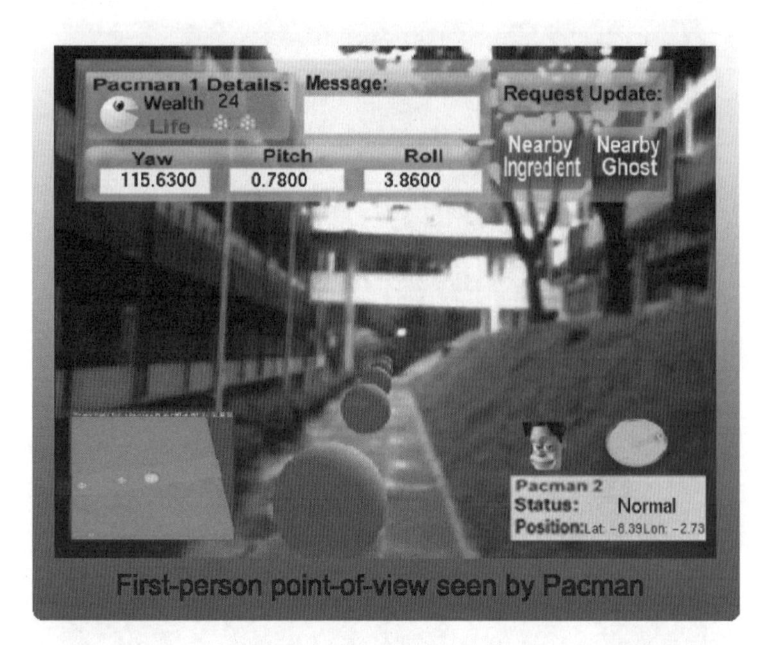

Figure 3.22 Screenshot of Human Pacman showing first person view of game from head-mounted display. Reproduced by permission of Adrian Cheok.

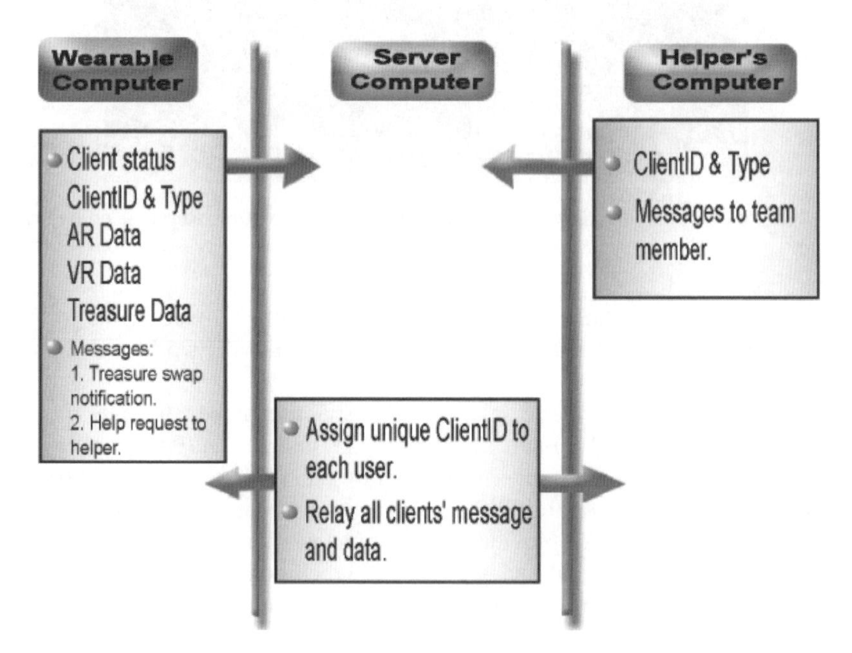

Figure 3.23 Information flow in Human Pacman. Reproduced by permission of Adrian Cheok.

Figure 3.24 ARQuake, with user wearing head-mounted display and holding a gun (a) and game scene depiction (b). Reproduced by permission of Wayne Piekarski.

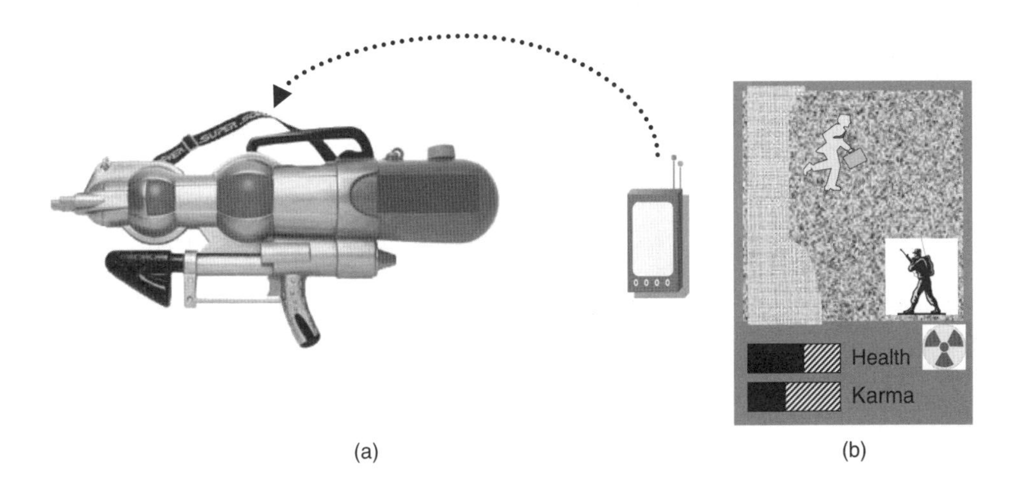

Figure 3.25 Real Tournament, showing super-soaker gun with attached IPAQ computer (a), and game scene depiction (b)

Another AR, location- action game uses mobile phones, Bluetooth network technology and high-precision location-aware hardware, called *BATS* [MSTM04]. The BATS mobile device has enough precision to infer user gestures such as picking virtual objects up or shooting (Figure 3.26). The mobile phone provides map information with the location of game objects to augment the reality of the physical walls and doors. A centralised computer that controls the game play also allows observers to monitor game play or even serve as commanders surveying the game battlefield.

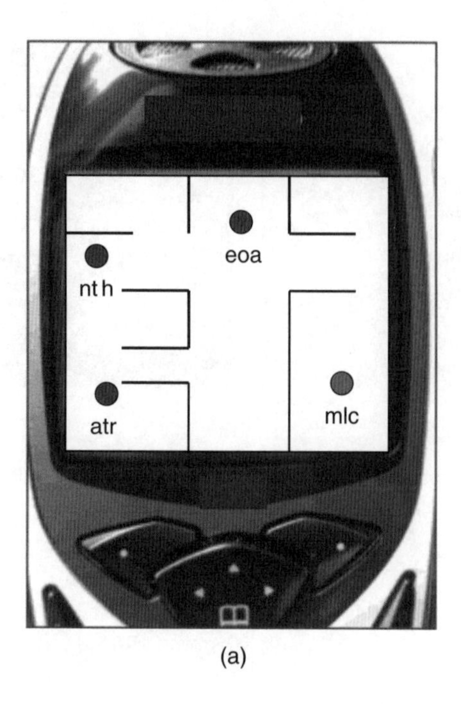

(a)

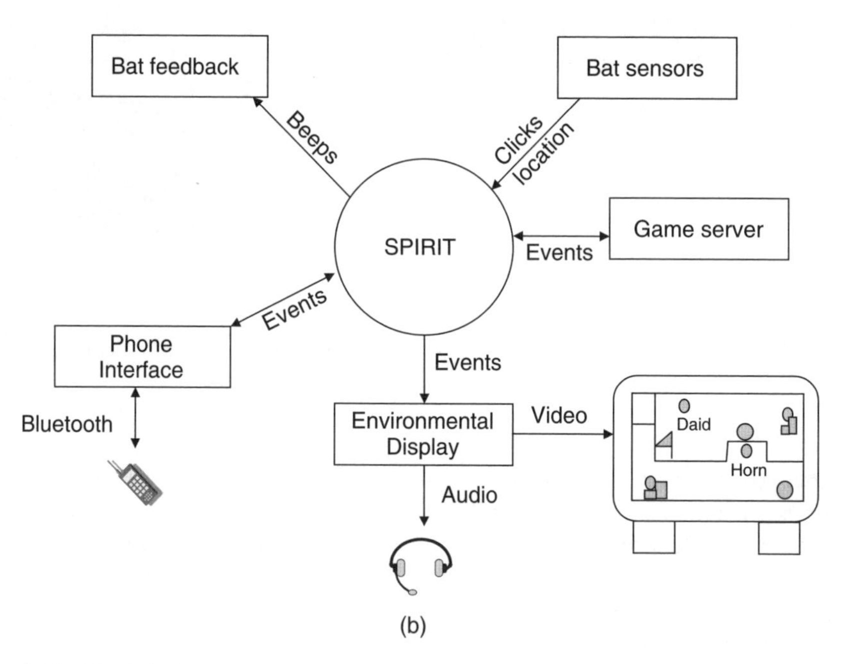

(b)

Figure 3.26 The BATS game, with a mobile phone providing the interface (a), and the game architecture (b)

AR provides a means of combining computer gameplay, physical exercise and social interaction. Beyond entertainment, more serious uses include the potential for military or disaster response training, allowing physical training exercises with computer augmenting physical reality with virtual conflicts.

3.5.4 Distributed Virtual Environments

Most computer games are in the virtual reality cluster, and within this cluster, virtual reality has produced three branches of research that have some relevance to computer games [SKH02]: *military simulations*, *Computer Supported Cooperative Work* (CSCW), and *virtual reality*.

(a) *Military simulations* have proceeded through *Distributed Interactive Simulation* (*DIS*), an IEEE standard for communication designed to allow networked simulators to interact through simulation using compliant architecture, modelling, protocols, standards and databases [Ney97]. A DIS system must be flexible and powerful enough to support up to hundreds or even thousands of simulators. The *Virtual Cockpit*, a tactics trainer for pilots, is an example of a typical vehicle that participates in DIS [MSA+94]. The Virtual Cockpit is a low-cost, manned flight simulator of an F-15E jet fighter built by the Air Force Institute of Technology. A soldier flies the Virtual Cockpit using the hands-on throttle and stick, while the interior and out-the-window views are viewed within a head-mounted display.

(b) *Computer Supported Cooperative Work* (CSCW) focuses on using computers to support the collaboration of users. An example of a CSCW application would be a shared whiteboard, where participants can read, edit, and annotate a shared document using a variety of media including text, pictures, video and voice. Efforts to combine CSCW with virtual reality has produced *collaborative virtual environments* (CVEs) [BGRP01] with network architectures to support their online use [SG01]. Successful CVE applications allow users to remotely control avatars to operate on a shared media, allowing 3-D editing, product development and even game design.

(c) *Distributed Virtual Environments* (DVEs) are technology-oriented environments for users to immerse themselves in the virtual world. An example of a DVE is a 'Cave' wherein users immerse themselves in an artificial 3-D world using various combinations of VR goggles, VR gloves and wands and various 3-D audio components [http://www.evl.uic.edu/pape/CAVE/, CSD+92]. DVEs concentrate on the player representation and the technology for interacting with the world. Typically, they only support participation by a few, physically local users, although architectures are being developed to allow larger-scale interaction by more participants [BWA96, FS98].

References

[BGR+98] S. Benford, C. Greenhalgh, G. Reynard, C. Brown and B. Koleva, "Understanding and constructing shared spaces with mixed-reality boundaries", *ACM Transactions on Computer-Human Interaction (TOCHI)*, Vol. 5, No. 3, pp. 185–223, 1998.

[BGRP01] S. Benford, C. Greenhalgh, T. Rodden and J. Pycock, "Collaborative virtual environments", *Communications of the ACM*, Vol. 44, No. 7, pp. 79–85, 2001.

[BWA96] J.W. Barrus, R.C. Waters and D.B. Anderson, "Locales: supporting large multiuser virtual environments", *IEEE Computer Graphics and Applications*, Vol. 16, No. 6, 1996.

[CFG+03] A. David Cheok, S. Wan Fong, K. Hwee Goh, X. Yang, W. Liu and F. Farzbiz, "Human Pacman: A Sensing-based Mobile Entertainment System with Ubiquitous Computing and Tangible Interaction", *Proceedings of ACM Network and System Support for Games Workshop (NetGames)*, Redwood City, CA, May 2003.

[CNET05] CNET Networks, Inc.,"Playstation3–Xbox 360: Tech Head to Head", 05/16/2005 [Online] http://hardware.gamespot.com/Story-ST-x-1985-x-x-x

[CSD+92] C. Cruz-Neira, D.J. Sandin, T.A. DeFanti, R. Kenyon and J. Hart, "The CAVE: audio visual experience automatic virtual environment", *Communications of the ACM*, Vol. 35, No. 6, pp. 65–72, 1992.

[FS98] E. Freecon and M. Stenius, "DIVE: A scalable network architecture for distributed virtual environments", *Distributed Systems Engineering*, Vol. 5, No. 3, pp. 91–100, 1998.

[Ger02] B. Geryk, "A History of Real-Time Strategy Games, Part I: 1989-1998", *Gamespot*, 2002, [Online]http://www.gamespot.com/gamespot/features/all/real_time.

[Ken03] S.L. Kent, "Making an MMOG for the Masses", GameSpy.com, October 10, 2003.

[Kus03] D. Kushner, *Masters of Doom*, Random House, 2003, ISBN: 0-375-50524-5.

[MADDEN05] Wikipedia,"Madden NFL", (as of 20 January 2006) [http://en.wikipedia.org/w/index.php?title=Madden_NFL&oldid=36020223

[MS94] M.J. Massimino and T.B. Sheridan, "Teleoperator Performance with Varying Force and Visual Feedback", *Journal of Human Factors*, Mark Vol. 36, No. 1, pp. 145–157, 1994.

[MSA+94] W. Dean McCarty, S. Sheasby, P. Amburn, M.R. Stytz and C. Switzer, "A Virtual Cockpit for a Distributed Interactive Simulation", *IEEE Computer Graphics and Applications*, Vol. 14, No. 1, pp. 49–54, January 1994.

[MSTM04] K. Mansley, D. Scott, A. Tse and A. Madhavapeddy, "Feedback, Latency, Accuracy: Exploiting Tradeoffs in Location-Aware Gaming", Proceedings of ACM Network and System Support for Games Workshop (NetGames), Portand, OG, 2004.

[NC04] J. Nichols and M. Claypool, "The Effects of Latency on Online Madden NFL Football", In Proceedings of the 14th ACM International Workshop on Network and Operating Systems Support for Digital Audio and Video (NOSSDAV), Kinsale, County Cork, Ireland, June 16–18, 2004.

[Ney97] D.L. Neyland, "Virtual Combat: A Guide to Distributed Interactive Simulation", *Stackpole Books*, Mechanicsburg, PA, 1997.

[PCS+06] L.S. Ping, A.D. Cheok, J.T. Kheng Soon, G.P. Lyn Debra, C.W. Jie, W. Chuang and F. Farzbiz, "A Mobile Pet Wearable Computer and Mixed Reality System for Human – Poultry Interaction through the Internet", Springer Journal of Personal and Ubiquitous Computing, ISSN: 1617-4909, (Paper) 1617-4917, (Online)03 November, 2005.

[PT02] W. Piekarski and B. Thomas, ARQuake: the outdoor augmented reality gaming system, *Communications of the ACM*, Vol. 45, No. 1, pp 36–38, 2002.

[QWORLD] Quake World.net, http://www.quakeworld.net/, Accessed 2006.

[SG01] S. Shirmohammadi and N.D. Georganas, "An End-to-end Communication Architecture for Collaborative Virtual Environments", *Computer Networks*, Vol. 35, No. 2–3, pp. 351–367, 2001.

[SKH02] J. Smed, T. Kaukoranta and H. Hakonen, Aspects of networking in multiplayer computer games, *The Electronic Library*, Vol. 20, No. 2, pp. 87–97, 2002.

[Tut04] W. Tuttle, "The Future of Live?", *GameSpy*, October 24, 2004, [Online] http://xbox.gamespy.com/articles/559/559846p1.html.

[Tys06] J. Tyson, "How Video Game Systems Work", HowStuffWorks, Inc, [Online] http://computer.howstuffworks.com/video-game5.htm

[Woo05] B.S. Woodcock. "An Analysis of MMOG Subscription Growth", *MMOGCHART.COM* 12.0 29 29 2004. January 1, 2005. Online: http://www.mmogchart.com/.

4

Basic Internet Architecture

Many design decisions and end-user experiences of multiplayer, networked games derive from the particular nature and characteristics of Internet Protocol (IP) networks. In this chapter we will cover the following core aspects of IP networking:

- 'Best effort' service
- IP addressing of hosts and other endpoints in the network
- Transport protocols – Transmission Control Protocol (TCP) and User Datagram Protocol (UDP)
- The difference between unicast, multicast, and broadcast communication
- Networks as meshes of routers and links
- Network hierarchies, address aggregation and shortest-path routing protocols
- Address management – Dynamic Host Configuration Protocol (DHCP), Network Address Translation (NAT) and the Domain Name System (DNS).

Feel free to skip this chapter if you already understand IP networking basics (such as IP addressing, subnets, prefixes, shortest-path routing, the role of routers and routing protocols). This chapter is primarily to refresh your memory and provide a backdrop for the interaction between IP network services and networked games.

We will illustrate IP networking principles with examples based on the current Internet's core technology, known as IP version 4 (IPv4) [RFC791]. We will review how IP networks come in a variety of sizes, the rationale behind IP addressing, the differences between unicast and multicast packet delivery, the roles of the TCP and UDP transport layer protocols, hierarchies in network routing, and shortest-path routing protocols. (We will not discuss an emerging new version known as IP version 6 (IPv6). IPv6 has broadly similar architectural characteristics and is not covered in this book. Even the most optimistic estimates do not see IPv6 being widely relevant to consumer-based networked games until 2010 or beyond.)

Figure 4.1 attempts to illustrate how end-user applications (such as our favourite networked games) and support services (such as DNS or DHCP, which are rarely exposed to the end user) are layered on top of the basic data transport services provided by an IP network. The *Internet Protocol* is so named because it hides the many underlying technologies that can make up an IP network (such as optical fibre links, microwave links,

Networking and Online Games: Understanding and Engineering Multiplayer Internet Games
Grenville Armitage, Mark Claypool, Philip Branch

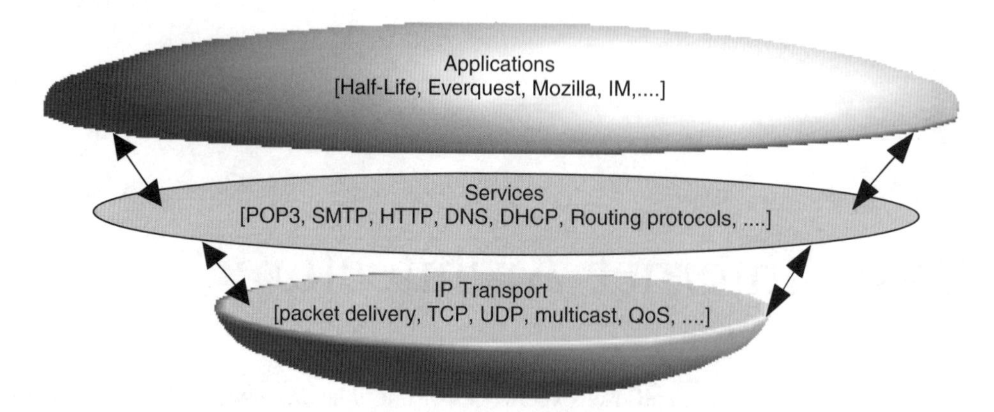

Figure 4.1 End-user applications and services utilise packet transport services provided by the IP network

ethernet Local Area Networks (LANs), 802.11/WiFi networks, cable modems and so on). IP provides a single, global addressing scheme across all the underlying technologies.

4.1 IP Networks as seen from the Edge

From an end user's perspective, there is a reasonable analogy between the postal service and an IP network. With traditional postal service we place letters into envelopes, address them to the final destination, and place them into a local post box. After that, we simply trust the postal service to transport our envelope to its destination in some reasonable time. We neither know nor care how the envelope gets to the destination, the delivery time is measured in days, and we accept that envelopes are sometimes lost. And we implement our own end-to-end strategies to confirm delivery (such as a phone call to the recipient some days later, or reposting a copy of the original letter every few days until the recipient responds). The postal service is a network, and its edges are the post-boxes and letter-boxes where we drop off and pick up our mail.

A simple way to view an IP network is as an opaque cloud with devices attached around the edges. These edge devices may be any piece of hardware (or software) that transmits or receives digital information. The primary goal of an IP network is to provide connectivity, that is, deliver data (in packets) from sources (who transmit packets) to destinations (who receive packets). Devices at the edges of IP networks typically act as both sources and receivers. IP edge devices (or endpoints) are identified with IP addresses, and endpoints are required to look after their own needs for reliable delivery of data. And finally, the IP network is presumed to be totally agnostic towards the actual contents of the packets each endpoint is sending.

Traditionally, IP networks provide few guarantees of timeliness or certainty in packet delivery – usually referred to as *best effort* service (although it might be more aptly considered a 'no guarantees' services). Milliseconds, hundreds of milliseconds, or seconds may elapse between the time a source injects (transmits) a packet into the network cloud, and the destination edge receives that same packet. Sometimes packets simply get lost inside the network and never arrive at all.

The time it takes for a packet to reach its destination is often referred to as *latency*. Short-term variation in this latency from one packet to the next is referred to as *jitter*. Packet loss is often described in terms of a *packet loss rate* – the probability of an IP packet being lost across a certain part of the IP network.

Despite the simplicity of best effort service, IP networks can be quite complex internally. In all but the most trivial networks, there will be multiple internal paths a packet may take between any given source and destination. The network must continually make internal choices as to which path is most optimal at any given time. This is the task of *routing protocols*, which we will discuss later in the chapter.

4.1.1 Endpoints and Addressing

IPv4 endpoints are identified with numerical 32-bit (4 byte) values, conventionally written in *dotted-quad* form – four decimal numbers (representing each of the four bytes making up the IP address) separated by periods. For example, the 32-bit binary address 1000 0000 0101 0000 1100 0101 0000 0111 is written as 128.80.195.7 in dotted-quad form. In theory, this allows for 2^{32} IPv4 addresses. Later in this chapter we will discuss why significantly fewer than 2^{32} IPv4 addresses are available in practice.

Endpoints are also often referred to as *hosts*, although a host (whether a Personal Computer (PC), game console with network port, or any device capable of attaching to a IP network) may have multiple interfaces to an IP network. A host with multiple interfaces (commonly referred to as a *multihomed host*) will have unique IP addresses on each of its interfaces to the IP network. Servers (e.g. web sites or game servers) and clients (e.g. someone running a browser on their home computer) are both hosts on the IP network.

Figure 4.2 shows a simple case where an IP network has three hosts attached, with IP addresses 136.80.1.2, 21.80.1.32, and 142.8.20.8. When host 136.80.1.2 wants to send an IP packet to another host (for example, host 142.8.20.8), it specifies this destination address 142.8.20.8 in the packet and passes the packet to the IP network. The IP network itself worries about how to locate and reach 142.8.20.8.

As a consequence of IP networking's hierarchical routing scheme (described later in this chapter), an IP address is tied closely to where the host is attached in the network.

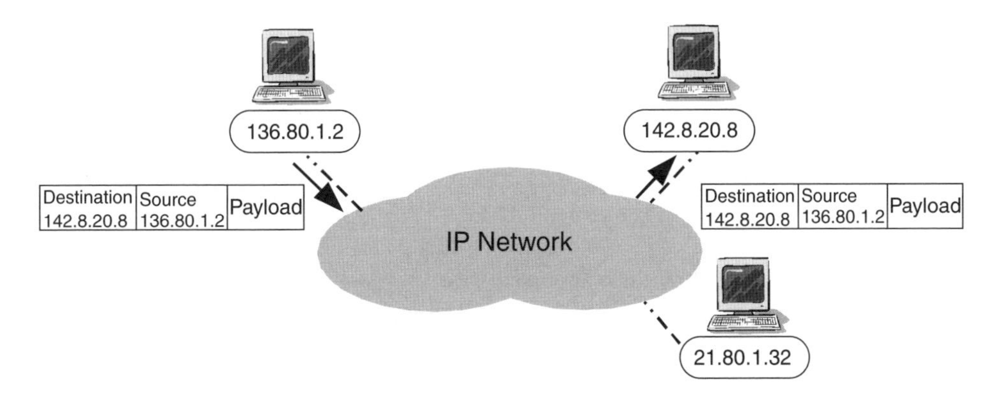

Figure 4.2 An IP Network – opaque cloud with attached devices – looks after getting packets to their destinations

In other words, an IP address represents both the identity of the attached host and the host's 'location' on the network. (This location is topological rather than geographical. It reflects where the host exists within the interconnections of IP networks and service providers that make up the Internet.)

IP addresses are closely related to, but not the same as, Fully Qualified Domain Names (FQDNs, or simply 'domain names'). Domain names (often imprecisely referred to as Internet addresses) are textual addresses of the form 'www.gamespy.com', 'www.freebsd.org' or 'www.bbc.co.uk'. Domain names must be resolved into IP addresses using the Domain Name System (DNS). Endpoint applications typically hide this translation step from the user, and use the resulting numeric IP address to establish communication with the intended destination. (We will discuss the DNS in greater detail later in this chapter.)

4.1.2 Layered Transport Services

Most game developers will utilise IP in conjunction with either the TCP [RFC793] or UDP [RFC768]. TCP and UDP are *transport* protocols, designed to provide another layer of abstraction on top of the IP layer's network service. Both TCP and UDP support the concurrent multiplexing of data from multiple applications onto a single stream of IP packets between two IP hosts. TCP additionally provides reliable delivery on top of the IP network's best effort service.

4.1.2.1 Transmission Control Protocol (TCP)

Early Internet applications – such as email, file transfer protocols and remote console login services – were sensitive to packet loss but relatively insensitive to timeliness (everything sent had to be received, but delays from tens of milliseconds to a few seconds were tolerable). The common end-to-end *transport* requirements of such applications (reliable ordered transfer of bytes from one endpoint to another) motivated development of TCP.

TCP sits immediately above the IP layer within a host (see Figure 4.3), and creates bidirectional paths (sometimes referred to as *TCP connections* or *TCP sessions*) between endpoints. An application's outbound data is broken up and transmitted inside TCP frames, which are themselves carried inside IP packets across the network to the destination. The

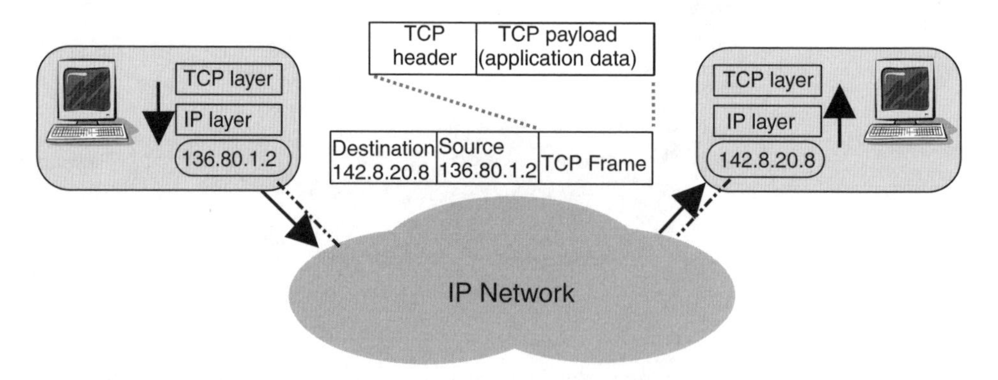

Figure 4.3 TCP runs transparently across the IP network

destination host's TCP layer explicitly acknowledges received TCP frames, enabling the transmitting TCP layer to detect when losses have occurred. Lost TCP frames are retransmitted until acknowledged by the destination, ultimately ensuring that the application's data is transferred with a high degree of reliability.

TCP uses *windowed flow control* to regulate how fast it transmits packets through the network. The current window size dictates how many unacknowledged packets may be in transit across the network at any given time. The source grows its window as packets are transmitted successfully, and shrinks its window when packet loss is detected (on the assumption that packets are only lost when the network is briefly overloaded). This regulates the bandwidth consumed by a TCP connection. Flow control and retransmission are handled independently in each direction.

Because TCP may keep retransmitting for many seconds when faced with repeated packet loss, the end application can experience unpredictable variations in latency (Figure 4.4). Thus, TCP is generally not suitable as the transport protocol for real-time messaging during game play of highly interactive networked games.

4.1.2.2 User Datagram Protocol (UDP)

UDP is a much simpler sibling of TCP, providing a connectionless, unreliable, datagram-oriented transport service for applications that do not require or desire the overhead of TCP's service. UDP imposes no flow control on packet transmission, and no packet loss detection or recovery. It is essentially a multiplexing layer sitting directly on top of IP's best effort service. As such an application using UDP will directly experience the latency, jitter and loss characteristics of the underlying IP network.

4.1.2.3 Multiplexing and Flows

Extending the postal service analogy a little further, while the IP address is analogous to a street address both TCP and UDP add the notion of *ports* – additional identification analogous to an apartment number or hotel room number. Each TCP or UDP frame carries two 16 bit *port numbers* to identify the source and destination of their frame within the context of a particular source or destination IP host. This allows multiplexing of different traffic streams between different applications residing on the same source and destination IP endpoints.

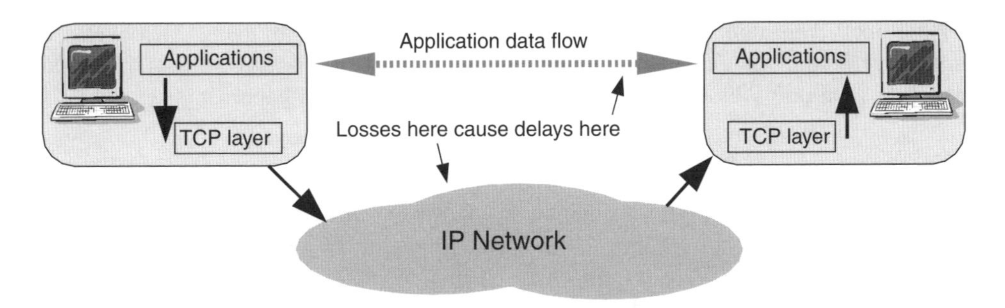

Figure 4.4 TCP converts IP layer packet loss into application layer delays

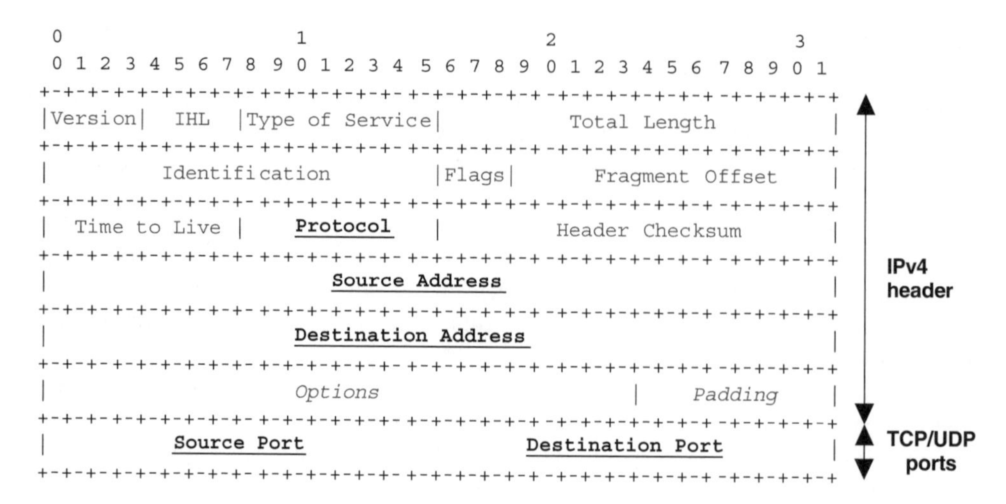

Figure 4.5 Header fields of interest in IPv4 packets

IP address and port number combinations are often written in the form '*ip-address:port*', with a ':' separating the address (either in dotted-quad or fully qualified domain name form) and the numerical port number.

Figure 4.5 shows the key fields of an IPv4 header and the first 32 bits of the TCP or UDP transport header. The protocol field specifies whether the IP packet carries TCP (protocol 6), UDP (protocol 17), or some other type of frame (discussed further in 'Directory of General Assigned Numbers' [IANAP]). The source and destination addresses identify a packet's source and destination host at the IP level.

Taken together, port numbers and IP addresses uniquely identify the source and destination applications that are generating and consuming the traffic. A sequence of packets exchanged between the same TCP or UDP ports on the same two endpoints is often referred to as an *application flow* (or just *flow*). Many applications use 'well-known' port numbers, often making it possible to infer the identity of an application from the source or destination port numbers. For example, the Simple Mail Transport Protocol (SMTP) typically uses TCP to port 25 on the mail server host [RFC2821], Quake III Arena servers default to using UDP port 27960, Half-Life 2 servers default to using UDP port 27015 and web servers typically respond to Hypertext Transport Protocol (HTTP) traffic on TCP port 80 [RFC2616].

Note that there are no rules preventing applications from using unconventional ports – we could, for example, just as easily run Quake III Arena on port 27015 and Half-Life 2 on port 27960, so long as everyone knows what is happening.

4.1.3 Unicast, Broadcast and Multicast

Sending a packet to a single destination is known as *unicast* transmission. Sending a packet to all destinations (within some specified region of the network) is known as *broadcast* transmission. Broadcasting may be implemented as multiple separate unicast transmissions, but this requires the source to actually know the IP addresses of all intended destinations. Usually the network supports broadcast natively – the source sends a single

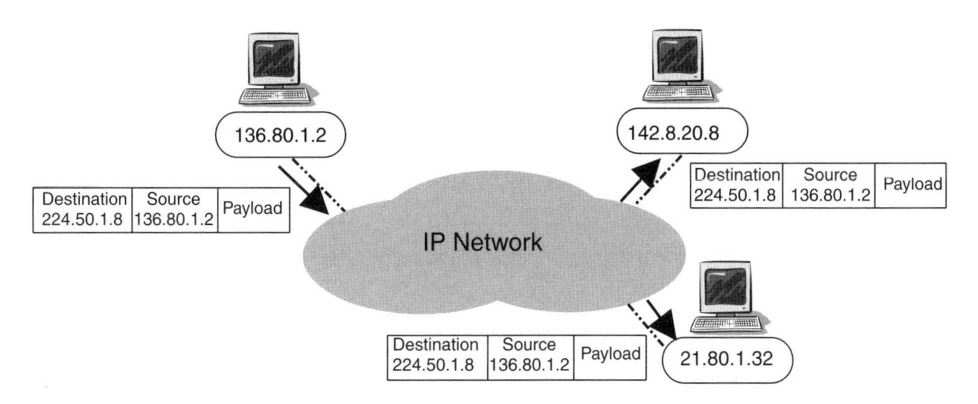

Figure 4.6 IP multicasting replicates a single packet to (potentially multiple) group members

packet into the network with a specific 'broadcast' destination address, and the network itself replicates the packet to all attached hosts within a restricted region.

A little-used alternative is *IP multicast* [RFC1112]. A source transmits one packet and the network itself delivers identical copies to multiple destinations (known as a *multicast group*, identified by special 'class D' IP destination addresses). Hosts explicitly inform the network when they wish to join or leave multicast groups. (Broadcast can be considered a special case of multicast, where every endpoint within a specific region of the network is automatically considered to be a group member.)

In IPv4, addresses in the range 224.0.0.0 to 239.255.255.255 are class D addresses, and represent multicast groups. Sources indirectly specify group members by using a class D address in their packet's destination address field.

Two attractive qualities of IP multicast are that a source does not need to track the multicast group members over time, and a source only sends one copy of each packet into the network. The network itself tracks group members and performs the necessary packet replications and deliveries.

Figure 4.6 shows a packet being sent to a group identified only by the destination address 224.50.1.8. Only the network is aware that the group includes endpoints 142.8.20.8 and 21.80.1.32. IP multicast is an 'any to many' service – a multicast group can have many members, and anyone can transmit to a multicast group from anywhere on the IP network (even if they are not a member of the group).

IP multicast holds some promise as a mechanism for efficiently delivering content, that is intended for concurrent delivery to multiple recipients. For example, replicating common game state across multiple clients or servers. Unicast requires a source to transmit its packets multiple times (once for each recipient), while multicast requires only one packet per update. However, because of the internal complexity required to support IP multicast there is little support in most public IP networks. This makes IP multicast difficult to use in networked games beyond specially constructed private networks.

4.2 Connectivity and Routing

From the game developer's perspective, it is often not necessary to understand the internal structure of IP networks. It is usually sufficient to comprehend the network's behaviour as seen from the edges. However, it is valuable to reflect on the internal details if you wish to more fully understand the origins of IP addressing schemes, latency, jitter, and packet loss.

An IP network is basically an arbitrary topology of interconnected links and routers. These terms are often thrown around casually, so we will define them here as follows:

- Links provide packet transport between routers.
- Routers are nodes in the topology, where packets may be forwarded from one link to another.

Upon receipt of an IP packet the router's primary job is to pick another link (the *next-hop* link) on which to forward (transmit) the packet, and then to do so as quickly as possible. Except for simple networks a router will usually have more than one possible choice of the next-hop link. Routers implement routing protocols to continuously exchange information with each other, subsequently learning the network's overall topology and agreeing on the appropriate next hops for all possible destinations.

This approach is known as *hop-by-hop forwarding*:

- An independent next-hop choice is made at each router.
- Each next-hop choice usually depends solely on the packet's destination address field.
- Routing protocols ensure that the network's routers agree on a coherent set of next-hop choices for all possible destinations.

Consider the network in Figure 4.7 where multiple paths exist between 136.80.1.2 and 21.80.1.32. Router R1 could send the packet to R2 or R3, both of which have the capability to forward the packet even closer to 21.80.1.32. In this example, R1 decides to use R2 as the next hop toward 21.80.1.32, and R2 has decided to use R5 as its next hop toward 21.80.1.32.

An IP network provides a *connectionless* service because it can transport IP packets from source to destination without any *a priori* end-user signalling. However, it is not *stateless*. The set of all source-to-destination paths currently considered optimal by the routing protocols is the state of the entire network.

In the rest of this section, we will look at how network hierarchies and address aggregation have been used to minimise the amount of state information that routing protocols need to handle. We will also touch on some routing protocols used in the Internet today.

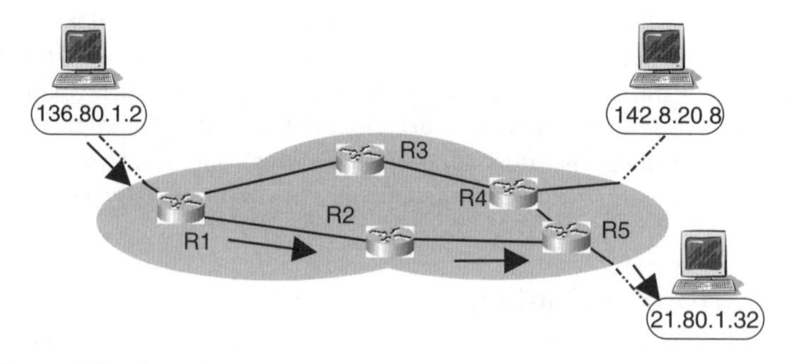

Figure 4.7 An arbitrary topology of routers may have multiple next hops

4.2.1 Hierarchy and Aggregation

The following issues are all closely related.

- IP address formats
- The association of IP addresses to endpoints
- How IP routing protocols establish appropriate paths?
- How routers make their next-hop forwarding decisions?

For small networks, it might seem reasonable for every router to simply know the identity and location of every endpoint. In practice, this approach is unworkable, as real networks may have thousands or tens of thousands of endpoints. Considering the many millions of hosts on the Internet itself, it is clearly impossible to expect routers (having only finite memory and processing capacity) to know all possible destinations.

The solution has been to introduce hierarchy into the IP address space – one that maps closely related IP addresses onto topologically localised sets of actual IP endpoints. This hierarchy allows routers to carry summarised information for regions of the network further away from them, and increasingly more detailed information for closer regions of the network. Hierarchy also creates sparseness of address allocation (consequently, far less than 2^{32} IP addresses can actually be allocated).

4.2.1.1 Class-Based Hierarchy

The IPv4 unicast address space was originally blocked into three classes – A, B and C (see Figure 4.8) [RFC791]. Specific combinations of an address' most significant 3 bits identified an addresses class. The next most significant 7, 14 or 21 bits of the IP address represented a *network number*. The Internet itself (at the time known as *ARPAnet*) was modelled as a backbone (a network of routers) with multiple independent networks directly attached. Each attached network was assigned a specific class A, B or C network number. Endpoints (hanging off each network) had their IP addresses constructed from their network's class bit(s), network number bits and a locally significant value for the remaining 24, 16 or 8 host bits. A router could easily determine which part of a packet's destination address represented the destination network, because the class of an IP address was encoded in the top 3 bits.

However, this class structure was particularly wasteful of address space. Many companies or institutions with more than 254 hosts had to obtain multiple class C networks (filling the backbones router tables) or a single class B (which would be barely utilised). In response, the Internet Engineering Task Force (IETF) developed Classless Inter Domain Routing (CIDR) in the early 1990s.

Class	Address Format in Binary	Networks	Hosts
A	0nnnnnnn.hhhhhhhh.hhhhhhhh.hhhhhhhh	2^7 nets	2^{24}hosts
B	10nnnnnn.nnnnnnnn.hhhhhhhh.hhhhhhhh	2^{14} nets	2^{16} hosts
C	110nnnnn.nnnnnnnn.nnnnnnnn.hhhhhhhh	2^{21} nets	2^8 hosts

Figure 4.8 Early IPv4 space divided into fixed-size classes

4.2.1.2 Classless Inter Domain Routing

CIDR replaced the previous A, B and C class rules (hence, *classless*) [RFC1519] with a flexible value/prefix-size pair scheme for identifying networks – the network number is encoded in the top bits of a 32-bit value, and the number of valid bits in the network number indicated by an integer prefix-size (Figure 4.9). Figure 4.9 shows that, in general, a prefix size of X results in a network that can theoretically contain up to $2^{(32-X)}$ endpoints.

A key benefit of CIDR was that variably sized networks could now be built from the old class C space. For example, 192.80.192/22 represents a single network with a 22-bit prefix and a network number of 192.80.192 – equivalent to four contiguous class C networks (192.80.192.*, 192.80.193.*, 192.80.194.* and 192.80.195.*, where '*' represents any number between 0 and 255). In other words, it represents a single '/22' network prefix in the backbone routers rather than four class C prefixes.

In the absence of CIDR, the last class B address would have been assigned in early 1994. CIDR significantly slowed the growth rate of the backbone routing tables, and increased the density with which IP addresses could be packed into a 32-bit field.

4.2.1.3 Subnetting

Creating a single network from multiple old class C networks is known as *supernetting*. The reverse, creating hierarchy within individual networks, is known as *subnetting*. Groups of endpoints may be aggregated into subnetworks (commonly referred to as *subnets*) if they are topologically localised within the scope of a larger network. Individual subnets contain endpoints whose addresses all fall under a common prefix (or *subnet mask*), a prefix that is itself a subset of the class or CIDR prefix assigned to the network of which they are a part. Subnets are networks within networks that can be described by a longer (that is, more precise) prefix or mask than the one that describes the network itself.

IP subnets are the lowest level of the IP routing and addressing hierarchy. Routing protocols do not concern themselves with local details within subnets. In all except the most simplistic network topologies, routers are needed in order to forward packets between subnets.

Layer 2 links between routers, such as Ethernet or similar LANs, are also often referred to as *subnets*. However, while multiple IP subnets may run over a single link, an IP subnet cannot (by definition) span more than one link without an intervening router.

Consider Figure 4.10, where Network 1 is made up of two internal subnets. The network's public identity (as advertised to the IP backbone's routers) is 128.80.0.0/16. Internally, Network 1 has two subnets – each with a longer, more precise 24-bit prefix (a subnet mask of 255.255.255.0). Subnet 1 covers all addresses in the range 128.80.1.0 to 128.80.1.255, whereas subnet 2 covers addresses in the range 128.80.9.0 to 128.80.9.255.

Subnet 1 and 2 may be geographically separate from each other yet owned by a common administrative entity (for example, a large company). Router R1 only advertises a

```
Address Format (binary)                              Networks and Hosts
nnnnnnnn.nnnnnnnn.nnnhhhhh.hhhhhhhh
|< - - - X - - - - >|                                2X nets, 2(32-X)
   hosts
```

Figure 4.9 CIDR relaxes the Network Prefix Lengths

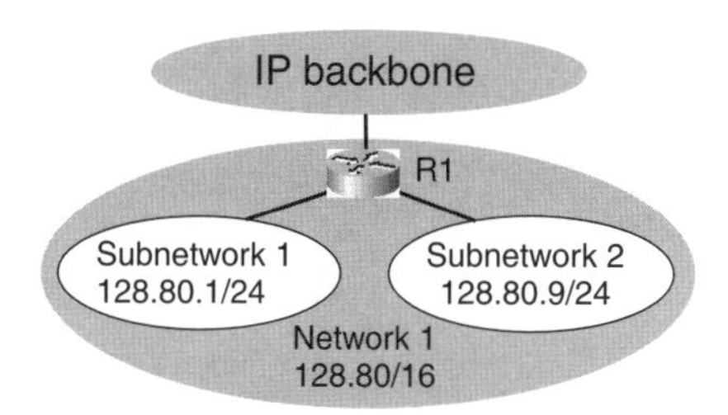

Figure 4.10 Subnetting allows aggregation within a network

single prefix (128.80.0.0/16) to the outside world, and takes care of forwarding packets to whatever subnets have been internally carved from the 128.80.0.0/16 address space.

Subnets may themselves be internally subnetted, with increasingly longer prefixes. Taken to an extreme, a subnet may map directly to a single link and have only two members (the IP interfaces at either end of the link).

The IPv4 address 255.255.255.255 is a special address meaning 'broadcast to all hosts on the local subnet'. Packets to 255.255.255.255 are never forwarded beyond the IP subnet on which they originate. A more general form, known as the *directed broadcast* address, is constructed by setting the host part of an IP address to ones. For example, you could transmit a packet to members of subnet 128.80.1.0/24 by using a destination address of 128.80.1.255. Because of the potential for remotely triggered mischief, routers are often set to filter out directed broadcast packets.

4.2.2 Routing Protocols

Network topologies change frequently, may be due to human interventions or the usual unpredictable failures that bedevil any large-scale system. Routing protocols must perform a number of tasks such as the following in a timely manner:

- Dynamically discover a network's topology, and track the topology changes that occur from time to time.
- Build shortest-path forwarding trees.
- Handle summarised information about external networks, possibly using different metrics to those used in the local network.

The Internet uses distributed routing protocols, which push topology discovery and route calculation processes out into every router. Since the processing load is shared across all routers, sections of the network can continue to adapt locally to changing conditions even if they become isolated from the rest of their network.

Figure 4.11 illustrates how every router participates both in forwarding packets (on the basis of previously calculated rules) and in performing distributed routing calculations (updating the forwarding rules as necessary).

The detailed art of IP routing is beyond the scope of this book, so we will only briefly summarise a few routing protocols used in the Internet.

4.2.2.1 Shortest-Path Routing

When multiple paths exist between a source and a destination, IP networks use *shortest-path* routing to pick one particular path. The 'length' of a path is typically measured in

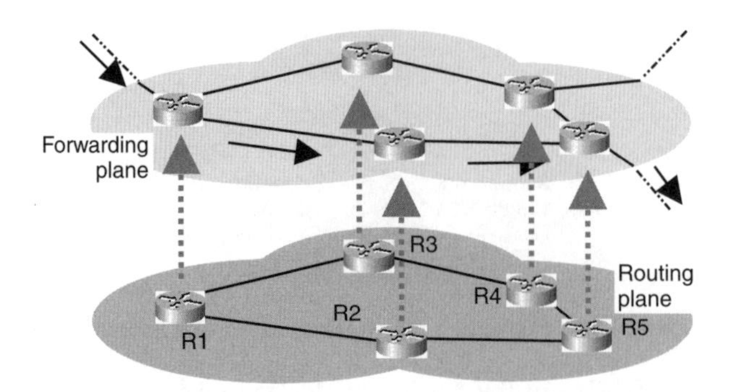

Figure 4.11 IP routing conceptually consists of separate forwarding and routing functions within each router

terms of hops (the number of routers or links through which the packet passes) but may be defined using any metric desired by the network operator. The path with the lowest 'sum of metrics over the entire path' is the shortest path. (for example, in Figure 4.7 the path through R1, R2 and R5 is the shortest path from 136.80.1.2 to 21.80.1.32 when measured by number of router hops.)

Metrics may reflect physical characteristics such as available bandwidth (lower weighting typically given to links with more bandwidth), link delays (higher weighting typically given to links with higher delay) and link costs; or weights may simply represent the administrator's relative preference for traffic to be on a particular link.

The results of a router's shortest-path calculations are stored as a set of forwarding rules in a *forwarding table*, sometimes also referred to as a *Forwarding Information Base* (FIB). Forwarding rules specify the appropriate next-hop destinations for packets matching various combinations of network/prefix pairs.

To ensure that routers always utilise the most precisely specified path, they are required to implement a *longest prefix match* when forwarding packets. In essence, the forwarding table's entries must be searched for the entry with the longest prefix that matches a packet's destination. The entry thus discovered is the correct next hop.

The routing protocol may also choose to use (or be required to account for) two extreme network/prefix pairs – *default routes* and *host routes*. Default routes are represented by the network/prefix 0.0.0.0/0 – a guaranteed match to any IP address. Because the prefix length is zero, this route is the last entry in a router's forwarding table.

Default routes are the ultimate in aggregation – if there is only one next-hop link out of the local network, a default route entry can point to that link (instead of having explicit forwarding rules for all the network/prefix pairs that can be reached in the world outside the local network). For example, in Figure 4.7 router R1 would have specific routes pointing into Network 1 for destinations under 128.80/16, and a default route entry pointing out toward the IP Backbone.

Host routes are represented by the network/prefix w.x.y.z/32 – a rule that only matches packets specifically destined for endpoint w.x.y.z. Host routes are discouraged because they are very difficult to aggregate and therefore can consume disproportionate amounts of memory resources in routers throughout the network.

Each destination prefix (whether a network, subnet, or actual host) known to the local network's routing protocol is said to be the root of its own particular *shortest-path tree*.

The tree has branches passing through every router in the network, although not all links in the network are branches on every shortest-path tree. No matter where a packet appears within the network, the packet will find itself on a branch of a shortest-path tree leading toward its desired destination.

The challenge for a dynamic IP routing protocol is to keep these shortest-path trees current in the presence of router failures, link failures or deliberate modifications to the network's topology – link failures usually require recalculating the shortest-path trees for many destination prefixes.

4.2.2.2 Autonomous Systems (ASs)

IP networks contain one additional component to their hierarchy – the Autonomous System (AS). An AS is defined loosely as a self-contained, independently administered network or internally connected set of networks. Larger networks, including the Internet itself, can be viewed as an arbitrary topology of interconnected ASs.

The AS exists to provide a bounded scope over which any given routing protocol must track internal topology. Within an AS, routing is managed by Interior Gateway Protocols (IGP, gateway being an old name for routers). Routing of traffic between ASs is managed by an Exterior Gateway Protocol (EGP). The IGP can focus on specific and detailed routes and destinations within the AS, while the EGP deals with summarised information about the AS and networks that can be reached through the AS.

4.2.2.3 Interior Gateway Protocols

Routing protocols that operate within ASs are called *IGPs. Distance Vector* (DV) and *Link State* (LS) algorithms have both been used as the basis of IGPs. DV algorithms tend to be simpler to implement, while LS algorithms are more robust when faced with regular topology changes within a network.

DV algorithms require each router to advertise to its neighbours information about the relative distance to each network the router knows (a vector of distances). A router may receive multiple advertisements for the same network X, each from a different neighbour, in which case the router remembers the advertisement with the lowest distance. The neighbour advertising the lowest distance toward X becomes the next hop for packets heading to any destination within network X. The advertising process (typically with intervals in the tens to hundreds of seconds) ensures that information about new networks, or new distances to existing networks, ripples out across the local network whenever changes occur.

LS algorithms distribute maps of the local network's entire topology (along with the state and metrics of all the links in the topology). The maps are distributed by flooding LS advertisements, whereby each router informs its neighbours about sections of the network topology that the local router knows about. When a state change occurs (for example, a link goes up or down, or a new route is associated with an existing link), the new LS information is flooded across the local network to ensure that all router's have up-to-date LS maps.

Each router then uses LS maps to locally calculate shortest-path trees to all listed destination networks, and hence determine the appropriate next hops out of the router itself. Because the next-hop calculations are based on complete knowledge of the network's state, every router can be expected to agree on the shortest-path trees.

Although DV protocols are simple to describe and implement, a network's shortest-path trees can get tangled up in transient loops while DV algorithms converge after topology changes. (Slow convergence is a fundamental limitation of any scheme in which the local router has only second hand, interpreted information about the nature of the network beyond the router's local interfaces.)

LS protocols are more complex than DV protocols. They contain two separate functions – maintenance of a distributed LS database and stand-alone shortest-tree calculation. The shortest-path trees can be assured to be loop free almost immediately after any LS changes occur and the information is flooded throughout the network. However, the network-wide flooding of state changes also limits the scalability of LS-based networks.

Two examples of DV IGPs are the Routing Information Protocol (RIPv2) [RFC2453] and IGRP (a proprietary IGP from Cisco Systems, which became Enhanced IGRP, EIGRP). Two examples of LS IGPs are Open Shortest Path First (OSPF) [RFC2328] and Intermediate System to Intermediate System (IS–IS).

4.2.2.4 Exterior Gateway Protocols

Border Gateway Protocol (BGP) version 4 is the standard EGP used in the Internet today [RFC1771]. BGP-4's primary role is to distribute information between ASs indicating where all the constituent networks are located. Every AS has one or more routers that interface to a peer AS – these are the *border routers* for the AS. Each border router runs an instance of BGP-4, enabling them to distribute to their neighbouring ASs information about the reachable networks within the local AS.

BGP-4 is a *path vector* protocol, which borrows a number of key DV concepts. In path vector, each border router advertises not only the existence of a path to particular networks (*reachability*) but also the list of ASs through which the path passes. Any given border router can confirm that an advertisement for a given network is loop free if the border router's own AS number does not already appear in the path vector. After an advertisement is accepted, the local border router inserts its own AS number into the path vector before readvertising the reachability information to its neighbours.

It is beyond the scope of this book to describe BGP's mechanisms to control the scope of reachability advertisements, support relative priorities between different inter-AS paths, and support policies that may restrict the ASs through which certain traffic can be routed.

4.2.2.5 Backbones and Routing Policies

There is no single backbone in today's Internet. Instead, we have a number of peer backbones. Top-level backbones typically interconnect only at a few geographically diverse points – Network Access Points (NAPs) or Internet exchanges (IX) – to ensure that any point on the Internet can connect to any other. Backbones interconnect at multiple points, providing redundancy against failure and potentially shortening many end-to-end paths.

However, political, geographical and/or commercial reasons mean that not all backbones wish to directly interconnect, even if physically possible. BGP-4 allows operators to constrain the advertisement of AS reachability in accordance with *routing policies* that reflect each operator's political or business agreements. As a result, IP paths may be convoluted simply because the source and destination are connected to different backbones that have no agreement to directly exchange traffic. Two geographically close sites (for example, London and Paris, Sydney and Melbourne, or Los Angeles and San Francisco) might find themselves communicating over paths that loop through New York, Tokyo, or Amsterdam depending on their choice of backbone provider and where the backbone providers interconnect.

4.2.3 Per-hop Packet Transport

This section reviews how an individual IP interface (whether on a router or an endpoint) uses the services of an underlying link to get IP packets to the appropriate next hop.

4.2.3.1 Link Layer Networks

For most of this chapter links have been treated as simple, point-to-point paths with only two interfaces attached. In reality, links are often networks in their own right. LAN (such as Ethernet) and wide area networks (such as frame relay) are examples of the link layers that support multiple attached devices. Devices attached to a link layer network may support IP, some other services, or a mixture of both.

Two addresses are associated with an IP interface attached to a given link layer network:

- The interface's IP address (representing the interface's identity in the IP topology).
- The interface's link layer address (representing the interface's specific identity in the context of the underlying link layer network).

In general, the link layer network is unaware of the IP address assigned to any given IP interface attached to the link. An IP packet's next hop (expressed as an IP address in a router's forwarding rules) must be translated to a link layer address before packet transmission can occur across the link.

Consider Figure 4.12, where a single Ethernet LAN [8023] has three attached interfaces, belonging either to routers or hosts on the 136.80.1/24 subnet. At the IP level, Interface 1 is known as 136.80.1.2, Interface 2 is known as 136.80.1.5, and Interface 3 is known as 136.80.1.9. Yet the Ethernet LAN only knows these interfaces by their 48-bit (6 byte) 'Media Access Control' (MAC) addresses (in this example MAC.1, MAC.2, and MAC.3, respectively).

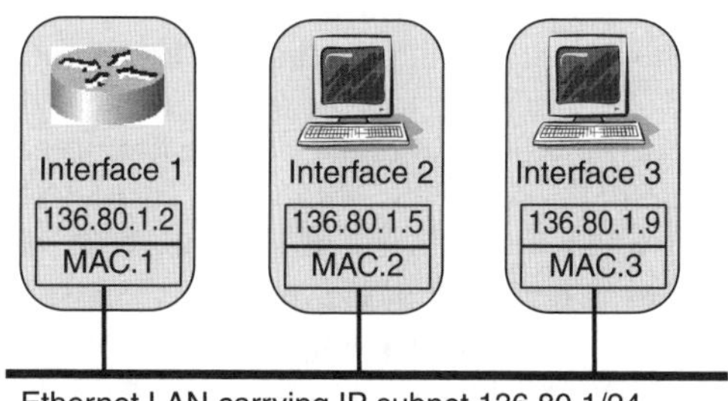

Figure 4.12 Each interface on an Ethernet LAN has both IP and Ethernet addresses

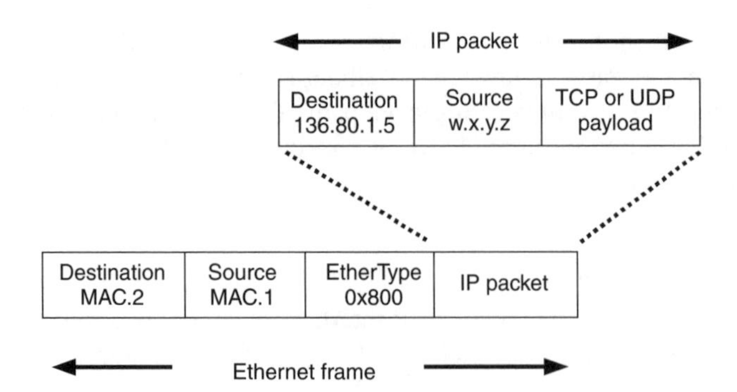

Figure 4.13 IP packets from the router at 136.80.1.2 to the host at 136.80.1.5 are encapsulated for transmission between Ethernet addresses MAC.1 and MAC.2

For example, if Interface 1 belonged to a router whose forwarding rules had just decided 'send this packet to the interface identified as 136.80.1.5', the following two steps would be executed

- A mapping would be established from 136.80.1.5 to MAC.2, the link layer address of Interface 2.
- The IP packet would be sent from MAC.1 to MAC.2 inside a suitably constructed Ethernet frame.

Figure 4.13 shows a highly simplified picture of how an IP packet from outside the 136.186.1/24 subnet would be encapsulated inside an Ethernet frame to be sent from Interface 1 to Interface 2. Ethernet frames carry an ethernet protocol type code – EtherType – of 0x800 to identify the payload as an IP packet (or more precisely, an IPv4 packet).

4.2.3.2 Address Resolution

Next-hop IP addresses are mapped to link layer addresses in a process referred to as *address resolution*. Address resolution may occur using information that is manually configured or is dynamically discover on-demand.

Manual configuration is unwieldy in all but the simplest of static network configurations. Most routers and hosts implement a dynamic Address Resolution Protocol (ARP) to identify what link layer address is associated with a particular IP address. ARP allows IP interfaces to move from one link layer interface to another without manually reconfiguring all the other interfaces attached to the link. This can be useful when, for example, an Ethernet card is replaced on a host or router – the Ethernet address changes, and the dynamic ARP process will ensure other interfaces on the link soon learn the new mapping. Figure 4.14 attempts to capture how ARP is both a peer of, and a service for, the IP layer.

Interfaces keep current address mappings in a local cache, an *ARP table*, which is searched each time a packet is transmitted. If a mapping exists, it is used. Otherwise the ARP is executed to discover the desired mapping. To ensure that old or incorrect mappings are regularly refreshed, cached ARP table entries are deleted after a period of time.

Each link layer technology has its own ARP mechanism. Some examples are ARP for IP over FDDI [RFC1390] and IP over ATM [RFC2225]. Perhaps the longest serving example

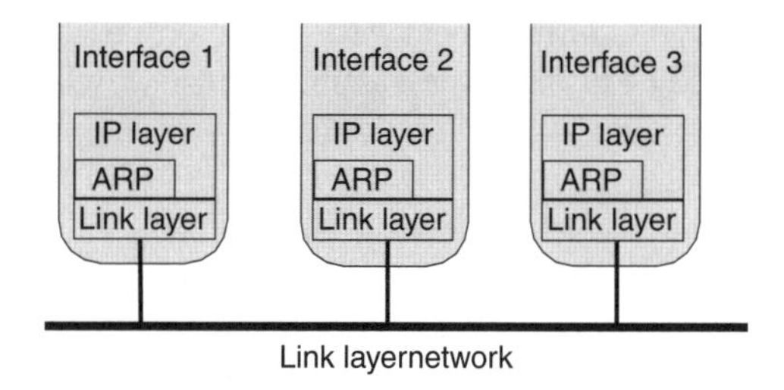

Figure 4.14 ARP is both peer of, and service for, the IP layer. Both sit over the Link Layer

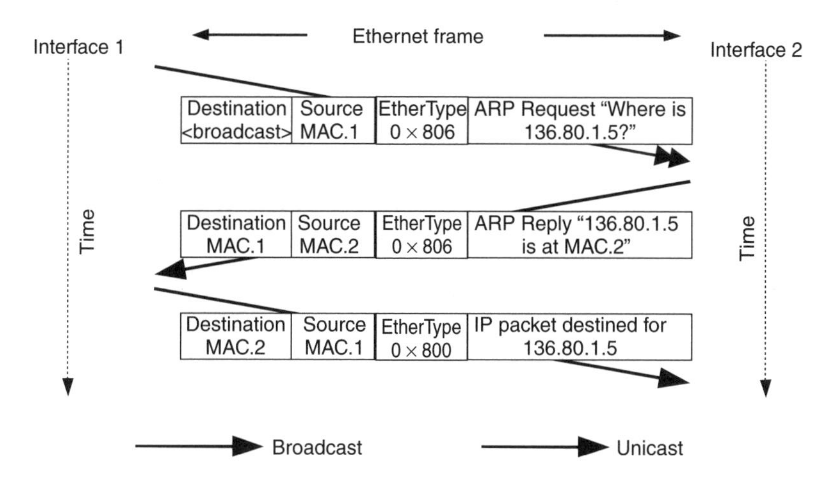

Figure 4.15 Frame sequence when initially sending an IP packet from Interface 1 to Interface 2

is ARP for Ethernet [RFC826], usually just referred to as *ARP*. When an IP address cannot be located in the local interface's ARP table, the interface issues a broadcast *ARP Request* on the LAN, essentially asking anyone else if they know the mapping. (Transmitting the ARP Request to the Ethernet broadcast address ensures that all attached devices are reached without the local interface needing to know who is, or is not, attached to the LAN at any given time.) Usually, the target (the interface whose IP address is being queried) will respond with a unicast *ARP Reply* containing the requested IP to Ethernet address mapping.

Figure 4.15 shows the packet exchange that would have occurred in the example of Figure 4.12 if Interface 1's local ARP table did not have a mapping for 136.80.1.5. (An EtherType 0x806 indicates an ARP Request or Reply, while EtherType of 0×800 indicates an IPv4 packet.) The sequence would be as follows:

1. Interface 1 transmits a broadcast ARP Request for 136.80.1.5.
2. Interface 2 unicasts back an ARP Reply (it knows the Ethernet address of Interface 1 from the initial ARP Request)
3. Interface 1 unicasts the IP packet to Interface 2.

For the curious reader: If you are running Windows XP or similar, entering *arp -a* in a console window will show the current local ARP cache entries. Under many versions of Unix (for example, FreeBSD or Linux) the command *arp -an* will show the current local ARP cache entries.

4.2.3.3 Time to Live

Transient errors in router forwarding tables can sometimes create loops in the IP network, known as *routing loops*. Routing loops tend to occur shortly after topology changes, while the network's routing protocol converges on a new set of shortest-path trees. Routing loops often act like black holes in the network – packets head into the region of the routing loop, and then get stuck, consuming bandwidth as they circulate. Under extreme circumstances, the routing loop can disrupt the routing protocol itself, by saturating links carrying routing protocol update messages.

To prevent endless looping, IP packets carry an 8-bit *Time to Live* (TTL) field (see Figure 4.5). In practice, the TTL represents 'hops to live' – a limit on the maximum number of router hops a packet can traverse before it expires in transit. A packet's TTL field is set to a nonzero value by the source, and is decremented by one every time the packet passes through a router. The packet is discarded when its TTL field is decremented to zero (whether or not it has reached its final destination).

If the source sets the TTL too low, some distant regions of the Internet may become unreachable. (Indeed, there were examples of this occurring with a popular PC operating system's default TTL in the early 1990s as the Internet became more topologically convoluted.) Setting the TTL too high increases the potential disruption a source's packets can cause during routing loops (by increasing the length of time the packets stay in transit around the loop). Many operating systems today set their initial TTL to 64 or some multiple of 32 above that (more by historical quirk than any particular mandatory requirement).

4.2.3.4 Maximum Transmission Units and IP Fragmentation

Although in principle, IP packets may be as large as 64 K bytes, most link layer technologies impose a substantially smaller limit on link level frame size. For example, Ethernet imposes a limit of 1500 bytes on the size of IP packets that can be carried in an Ethernet frame.

The underlying link layer's frame size limit is reflected at the IP layer by a parameter known as the *Maximum Transmission Unit* (MTU). When forwarding an IP packet larger than the link's MTU, IP interfaces must perform *IP fragmentation* – chopping (fragmenting) the IP packet up into a sequence of smaller IP packets that all fit under the MTU limit. IP fragmentation occurs underneath the TCP or UDP layer, which allows the source UDP- or TCP-based applications to be unaware of the actual MTUs of links along the path to the destination. The ultimate destination is responsible for reassembling the fragments into the original packet, and then treating the reassembled IP packet as though it had arrived in one piece.

IP fragmentation tends to occur when a packet's path originates on a link with a large MTU, and then at some point along the route passes across a link with a smaller MTU. It is not considered a good thing, as it creates less efficient data transfer along the path [RFC1191][RFC1981].

For the curious reader: Under many versions of Unix (for example, FreeBSD or Linux) the command *ifconfig* shows a variety of details about a host's currently attached link layers, including the currently assigned IP address(es), subnet mask(s), and MTU(s). If you are running Windows XP or something similar, the command *ipconfig /all* at a console window will also print out a variety of details about the host's current configuration.

4.2.3.5 First and Last Hops

Forwarding tables are not just for routers. Hosts also have a limited forwarding table that tells them the initial next hop (often referred to as the *first hop*) for outbound packets. The first hop will either be to a router (for traffic destined beyond the local subnet) or directly to a neighbour on the same subnet. In cases where the next hop goes directly to a neighbour on the same subnet it is referred to as the *last hop*. (The first and last hops may be one and the same in the case of communication between two hosts on the same subnet.)

Forwarding tables have special rules for subnets that are directly attached to one of the host's or router's IP interfaces. Rather than having a specific entry for every IP interface reachable on a local subnet, the 'next-hop' IP address is copied directly from the packet's destination address field. The ARP cache is then scanned for a match to this next-hop IP address, and the packet transmitted to the link layer destination found in the ARP cache.

Hosts typically only have a few entries in their forwarding table, for example, one entry for the directly attached subnet, and a default route (network/prefix 0.0.0.0/0) pointing to a router on the local subnet that provides access to the rest of the network. If a host has link interfaces to multiple IP subnets, it will have forwarding table entries for each directly attached subnet, and possibly multiple forwarding entries for nonlocal traffic.

For the curious reader: If you are running Windows XP or something similar, entering *route print* in a console window will show the current forwarding rules. Under many versions of Unix (for example, FreeBSD or Linux) the command *netstat -rn* will show the host's current forwarding rules.

4.2.3.6 Tunnels as Links

Links are simply mechanisms for getting an IP packet from one router to another. A link may even be an IP network in its own right. Transmission of IP packets within other IP packets is known as *IP tunnelling*, and the link is known as an *IP tunnel*. From the perspective of the outer IP packet, the packet being tunnelled is just another payload (as uninteresting as a TCP or a UDP frame). From the perspective of the tunnelled packet, the tunnel looks like another link layer. From an implementation perspective, an IP tunnel is a link layer where source and destination addresses also happen to be IP addresses.

The tunnel's endpoint is the IP interface identified by the destination IP address in the outer packet's header. When the outer IP packet reaches its destination, the original (inner) IP packet is extracted and processed as though it had arrived over a regular interface. The outer packet's IP Protocol Type identifies the payload as a tunnelled packet, for example, protocol type 4 indicates that the payload is an IPv4 packet [RFC2003]. Because a tunnel represents a single hop from the perspective of the tunnelled packet, its TTL is decremented by one (rather than the number of hops between the tunnel endpoints).

Tunnelling over an IPv4 network imposes an additional 20 bytes of overhead (the header of the outer IP packet) [RFC2003]. The MTU of the tunnel's virtual link is 20 bytes smaller than the smallest MTU along the outer packet's path. (RFC 2004 suggests a more efficient encapsulation mechanism incurring only 8 or 12 bytes of overhead but with some loss of generality, for example, fragmented IPv4 packets cannot be tunnelled [RFC2004]. For RFC2004-based tunnelling, the protocol type in the encapsulating/tunnelling header is 55.) When tunnelling over IPv6 networks, the MTU drops by at least 40 bytes (the size of the encapsulating IPv6 header) [RFC2473].

4.3 Address Management

Every IP interface needs an IP address, which raises some very real administrative issues when building big networks. There are two key aspects to address management:

- Establishing an IP subnet from which you can assign IP addresses
- Actually assigning individual IP addresses to interfaces.

In this section, we will review how address blocks are assigned to customer networks from blocks delegated to Internet Service Providers (ISPs); how NAT can be used to overcome limitations in ISP address assignments; how the DHCP simplifies address assignment to individual devices inside your networks and how the domain name service attempted to decouple the naming of endpoints from the addressing of endpoints.

4.3.1 Address Delegation and Assignment

To be part of the wider Internet you cannot pick a subnetwork number and prefix at random. You need IP addresses that are globally unique and routable on the Internet. Such addresses are typically obtained from your ISP, which assigns you addresses from larger blocks allocated to the ISP by regional *registries* around the world [RFC2050].

For example, ARIN (the American Registry for Internet Numbers) manages space under 204/8 (204.0.0.0/8) and a number of other large blocks of IPv4 address space, and APNIC (the Asia-Pacific Network Information Center) manages space under 218/8 and a number of other large blocks. An ISP who asks for space from APNIC will receive an allocation under 218/8 or one of APNIC's other address blocks. Regional registries develop their own policies for subdividing the address blocks they manage. Up-to-date information on registries and assignment policies can be found in the Internet Assigned Numbers Authority web site, http://www.iana.org.

Problems arise when you decide to change ISPs. You will usually be forced to adopt a new IP address space assigned by your new ISP (called *renumbering*). Renumbering of your network is usually required by ISPs because routing would become more convoluted if the address hierarchy was allowed to arbitrarily diverge from the hierarchy of actual connectivity among the service providers.

A number of IPv4 address blocks (10/8, 172.16/12 and 192.168/16) have been reserved for use in *private internets* [RFC1918]. These are useful when building IP networks that will never be connected to the Internet, or will be connected only in a very limited fashion. In principle, such IP networks could be built using any prefixes. However, using designated private IP address spaces helps administrators distinguish between internal and external hosts in cases where their network contains a mixture of internal and external connectivity.

Interesting problems arise when a previously private network wishes to connect to the Internet. These will be discussed in the following section, along with a currently popular solution – NAT.

4.3.2 Network Address Translation

NAT is used at the boundaries between IP networks, most often between private networks and the public Internet [RFC3022]. Fundamentally, it solves the problem of a private network whose internal IP address space does not map cleanly into an unused, publicly routable IP address space.

4.3.2.1 Pure NAT

Pure NAT is best explained with an example. Consider the situation in Figure 4.16. A company has a private network of 100 hosts, using addresses in the private range 192.168.0.1 to 192.168.0.100. At some point in time, the company wishes for all hosts on its internal IP network to access the Internet. The company contacts an ISP and is allocated a CIDR block of 256 addresses, perhaps 128.80.6/24. If there were only a router between the private network and the ISP, every host would need to be renumbered to a unique address in the 128.80.6/24 range.

However, NAT provides an alternative to renumbering. Basically, NAT dynamically modifies the source and destination addresses in packets as they are forwarded between the private network and the ISP.

For packets being transmitted out to the Internet, the steps are as follows:

- Source hosts use their own private address in the packet's source address field.
- Internal routing forwards packets to the NAT-enabled router linking the private network to the ISP.
- The NAT-enabled router swaps the source address in each packet with a source address taken from the publicly routable address space 128.80.6/24, and then forwards the packet to the ISP.

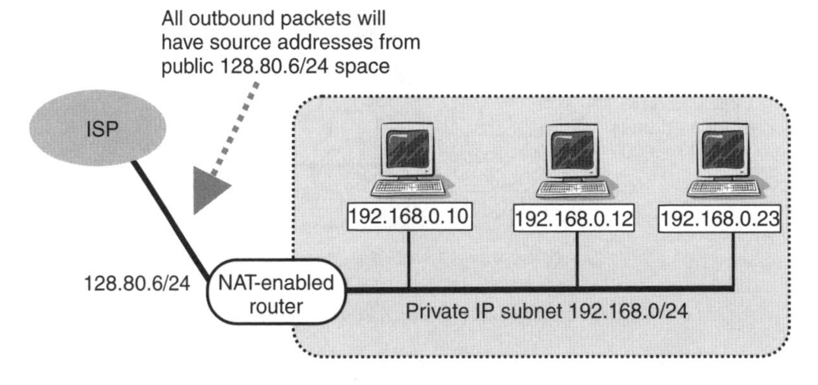

Figure 4.16 NAT helps map a private address space into the public address space

- The router remembers the address mapping it used, so it can reverse the process for inbound packets coming from the ISP.

For packets coming back in from the Internet, the reverse steps are as follows:

- A packet arrives at the NAT-enabled router, with a destination address in the 128.80.6/24 range.
- The NAT-enabled router looks up the mapping between 128.80.6/24 addresses and internal 192.168.0.*, and replaces the packet's destination address with the private IP address of the intended destination host.
- Internal routing (within the private network) forwards the modified IP packet to the correct destination.

A NAT-enabled router is generally free to use whatever address mapping schemes it chooses. For example, in Figure 4.16 the NAT-enabled router might choose to map internal address 192.168.0.10 to public address 128.80.6.20, or indeed any address in the 128.80.6/24 range. Mappings may be statically assigned, or dynamically generated on-demand. The only requirement is that mappings are unique – multiple private host addresses should never map to the same public IP address, and vice versa.

4.3.2.2 Network Address Port Translation

A common scenario for home networks is where the ISP (whether regular modem dial-up or a broadband service) charges additional monthly fees for a second or third public IP address. The solution is an extended version of NAT called *Network Address Port Translation* (NAPT) [RFC3022]. An NAPT-enabled router transparently makes multiple hosts on the private network appear to be a single host from the perspective of the public Internet.

NAPT extends NAT by additionally manipulating the port numbers of TCP and UDP traffic going in and out of the private network. Consider the scenario of Figure 4.17, where two hosts on a private LAN (192.168.0.12 and 192.168.0.13) are sharing a single public IP address (128.80.6.200).

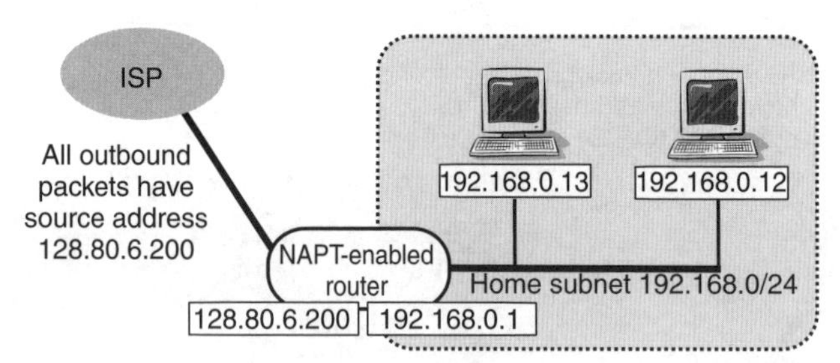

Figure 4.17 NAPT maps both addresses and TCP/UDP ports to share public IP addresses across multiple private hosts

Imagine that both hosts are engaged in separate TCP connections with other, unrelated hosts on the Internet. The host TCP connections originate from ⟨addr=192.168.0.12, port=W⟩ and ⟨addr=192.168.0.13, port=X⟩ respectively. Because there is no coordination between hosts when they choose their TCP source ports, there is no guarantee that W and X are different.

NAPT remaps both the source port and source address fields to ensure that the individual TCP connections appear unique on the public side of the router. In this example, outbound packets from ⟨addr=192.168.0.12, port=W⟩ are modified to appear as packets from ⟨addr=128.80.6.200, port=Y⟩ and forwarded to the ISP. Likewise, packets from ⟨addr=192.168.0.13, port=X⟩ are modified to appear as packets from ⟨addr=128.80.6.200, port=Z⟩, where Z and Y are guaranteed to be different. Packets coming back in from the Internet are modified with the reverse mappings before being forwarded onto the private network.

4.3.2.3 Convenience and Limitations

Both NAT and NAPT provide independence from the need to renumber when your public IP address(es) change. Only the NAT-enabled router needs to be aware of any change in the range of public IP addresses assigned to the company's network – the hosts remain unchanged. This makes it easy for small companies to change ISPs with minimal disruption of internal network operations. NAPT also allows multiple hosts on a home LAN to access the Internet while avoiding additional charges for more than one IP address.

Naturally, all this convenience comes with caveats [RFC2993]. NAT and NAPT break the transparency of TCP and UDP communication between hosts, and require special-case coding to handle other protocols. While hosts on the private network may initiate communication with anyone else on the Internet, the reverse is far more complex. Additional functionality is required in your NAT/NAPT router to enable hosts inside the private network to support 'well-known' servers visible to the rest of the Internet.

For example, imagine you have a small corporate site with 200 hosts and three of them want to run publicly accessible web servers. The default 'http://www.companyname.com' web address format actually implies that the web server is listening for HTTP traffic on TCP port 80. However, if you only have one public IP address, the NAPT router can only map inbound ⟨dst addr, port=80⟩ traffic to *one* of your internal hosts, not three. The second and third would-be web servers will either need to give up on their plans, merge with first machine, or configure the NAPT router to utilise nonstandard mappings (for example, mapping ports 8080 and 8081 to the second and third internal machines respectively, and giving external web addresses of the form 'http://www.companyname.com:8080' and 'http://www.companyname.com:8081' respectively).

Running game servers behind NAT/NAPT is similarly problematic. Many games require the server to register its IP address and port number with a master server (through which potential players find available game servers). But when sitting behind NAT/NAPT, inbound connections (e.g. from new players) are typically only allowed by the NAT/NAPT router if they correspond to a recently initiated outbound connection. But since players initiate contact with the game server, not the other way round, we have a dilemma. (For example, consider Figure 4.17 with a Quake III Arena server running on host 192.168.0.13 at port 27960. Further, assume the NAPT router is manually configured to map 128.80.6.200:28000 to 192.168.0.13:27960. The master server 'sees' the Quake III Arena server at 128.80.6.200:28000. However, without special configuration the NAT/NAPT router will not allow new players to actually connect through 128.80.6.200:28000 to the game server itself. We discuss this again in Chapter 12.)

NAT/NAPT has its admirers and detractors. Nevertheless, it does serve a purpose for private networks that cannot afford lots of public IP addresses or wish to avoid renumbering of their internal networks on a regular basis.

Consumer home routers/gateways invariably support some form of NAT/NAPT functionality. Demand is driven by the deployment of broadband IP access over Asymmetric Digital Subscriber Line (ADSL) or cable modem services, and the fact that many homes have multiple computers. Typically the home router has one Ethernet port to the ADSL modem or cable modem, and one or more Ethernet ports for the internal, home network. (To assist in address management of a small home network, many home routers also support the dynamic host configuration protocol described in the following section.)

4.3.3 Dynamic Host Configuration Protocol

The DHCP [RFC2131] automates the configuration of various fundamental parameters hosts need to know before they can become functional members of an IP network. For example, every host minimally needs to know the following:

- The host's own IP address
- The subnet mask for the subnet on which it sits
- The IP address of at least one router to be used as the default route for all traffic destined outside the local subnet.

Without these pieces of information, a host cannot properly set the source IP address of its outbound packets, cannot know if it is the destination of inbound unicast packets, and cannot build a basic forwarding table that differentiates between on-link and off-link next hops.

DHCP allows hosts to automatically establish the preceding information, and provides two key benefits:

- The need for manual intervention is minimised when installing and turning on new hosts.
- IP addresses can be *leased* for configurable periods of time to temporary hosts.

Minimising administrative burdens clearly saves money and time, and increases overall convenience. The benefits of dynamic address leasing become apparent in networks where not all hosts are attached and operational at the same time.

4.3.3.1 Configuring a Host

DHCP is a client–server protocol. Each host has a *DHCP client* embedded in it, and the local network has one or more nodes running *DHCP servers*. DHCP runs on top of UDP, which at first glance suggests a Catch-22 situation with the unconfigured IP interface. However, DHCP only requires that an unconfigured IP interface can transmit a broadcast packet (IP destination address '255.255.255.255') to all other IP interfaces on the local link.

A DHCP client solicits configuration information through a multistep process:

- First the client broadcasts a DHCP Discover message in a UDP packet to port 67, to identify its own link layer address and to elicit responses from any DHCP servers on the local network.
- One or more DHCP servers reply with DHCP Offer messages, containing an IP address, subnet mask, default router, and other optional information that the client may use to configure itself [RFC2132].
- The DHCP client then selects one of the servers, and negotiates confirmation of the configuration by sending back a DHCP Request to the selected server.
- The selected DHCP server replies with a DHCP Ack message, and the host begins operating as a functional member of the subnet to which it has been assigned.

Address administration is thus centralised to one or more DHCP servers.

4.3.3.2 Leasing Addresses

DHCP servers may be configured to allocate IP addresses in a number of ways:

- Static IP mappings based on *a priori* knowledge of the link layer addresses of hosts supposed to be on the managed subnet.
- Permanent mappings that are generated on-demand (the server learns client link layer addresses as clients announce themselves).
- Short-term leases, where the DHCP client is assigned an IP address for a fixed period of time after which the lease must be renewed or the address returned to the server's available address pool.

The third option is most useful where network access must be provided to a large group of transient hosts using a smaller pool of IP addresses. For example, consider a public access terminal centre at a university with 50 Ethernet ports into which students can plug their laptop computers. Many hundreds or thousands of students might use the centre over a week or a month. Thousands of IP addresses would be required for a static IP address assignment scheme, one for each student's laptop. On the contrary, leasing addresses for short periods of time means that a much smaller pool of IP addresses can serve the terminal centre's needs.

A DHCP server can specify lease times in the order of hours, days, or weeks with a minimum lease time of one hour. DHCP clients are informed of the lease time when they first receive their address assignment. As the lease nears expiration, DHCP clients are expected to repeat the Request/Ack sequence to renew their lease. The DHCP server usually allows clients to continue with their leases at renewal.

DHCP clients are also allowed to store their assigned IP address in long-term storage (battery-backed memory, or local disk drive) and request the same address again when it next starts up. If the address has not subsequently been issued to another client, DHCP servers typically allow a lease to be renewed after clients go through a complete restart.

4.3.4 Domain Name System

As noted earlier in this chapter, IP addresses are not the same as FQDNs (often referred to simply as *domain names*). Domain names are human-readable, text-form names that indirectly represent IP addresses.

The DNS is a distributed, automated, hierarchical look-up and address mapping service [RFC1591]. People typically use domain names to inform an application of a remote Internet destination, and the applications then use the DNS to perform on-demand mappings of domain names to IP addresses. This level of indirection allows consistent use of well-known domain names to identify hosts, while allowing a host's IP address to change over time (for whatever reason).

Two forms of hierarchy exist in the DNS – hierarchy in the structure of names themselves and a matching hierarchy in the distributed look-up mechanism.

4.3.4.1 Domain Name Hierarchy

Domain names are minimally of the form ⟨*name*⟩.⟨*tld*⟩ where ⟨tld⟩ specifies one of a handful of *Top Level Domains* (TLDs) and ⟨name⟩ is an identifier registered under the specified top-level domain. Examples of generic three letter TLDs (gTLDs) include com, edu, net, org, int, gov, and mil. Country code TLDs (ccTLDs) are constructed from standard ISO-3166 two letter 'country codes' (e.g. au, uk, fr, and so on) [ISO3166].

The nested hierarchy is read from right to left, and ⟨name⟩ may itself be broken up into multiple levels of subdomains. Some TLDs are relatively flat (for example, the 'com' TLD), with companies and originations around the world able to register second-level domains immediately under 'com'. Country code TLDs have varied underlying structures, sometimes replicating a few of the existing three letter TLDs as second-level domains (for example, Australia registers domain names under a range of second-level domains including 'com.au', 'edu.au', and so on.)

The hierarchical structure reflects the administrative hierarchy of authority associated with assigning names to IP addresses. For example, consider an address like 'mail.accounting.bigcorp.com'. The managers for 'com' have delegated all naming under 'bigcorp.com' to a second party (most likely the owners of 'BigCorp, Inc.'). BigCorp no doubt has various internal departments, including the Accounting department. Someone in the accounting department has been delegated authority for naming under 'accounting.bigcorp.com', and they have assigned a name for the mail server in the accounting department. Figure 4.18 represents the relationships between the subdomains discussed so far.

A domain name hierarchy is independent of the hierarchy of IP addresses and subnets discussed earlier in this chapter. For example, onemachine.bigcorp.com might well be on an entirely different IP subnet (indeed, even a different country) from othermachine.bigcorp.com.

Domain name registration has become a commercial business in its own right, and multiple *registrars* jointly manage different sections of the DNS. Up-to-date information on registrars and domain assignment policies can be found in the Internet Assigned Numbers Authority web site, http://www.iana.org.

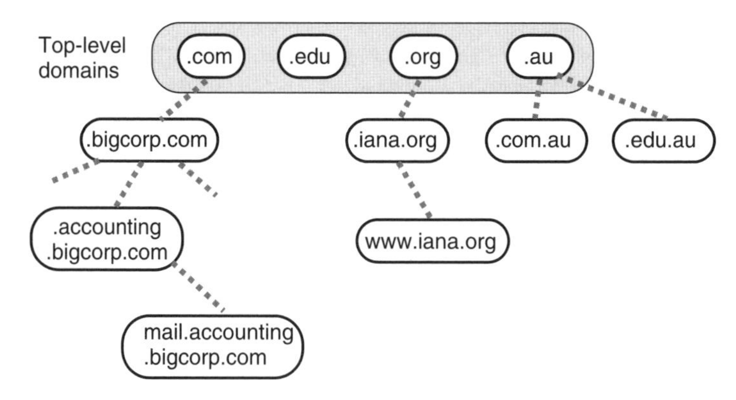

Figure 4.18 Hierarchy within domain name structure reflects a hierarchy of delegation authority

4.3.4.2 DNS Hierarchy

The hierarchy in a domain name essentially describes a search path across the distributed collection of *name servers* that together make up the Internet's DNS. Name servers are queried whenever a domain name needs to be resolved to an IP address. Certain name servers are responsible for being authoritative sources of information for particular domains or subdomains. The name servers ultimately responsible for each TLD are known as *root name servers.*

Before hosts can use the DNS they must be configured with the IP address of a local name server – the host's entry point into the DNS. The ISP or whoever supports your network typically provides the local name server. The name server's IP address is either manually configured into each host, or can be automatically configured (for example, DHCP provides an option for configuring the local name server's address [RFC2132]).

Local name servers are manually configured to know the IP address of at least one root name server, and possibly another name server further up the domain name tree. Name servers either answer queries with local knowledge, or seek out another name server who is responsible for mappings higher up the domain name hierarchy. Local knowledge is often held in a cache built from recent queries from other hosts – the cache allows rapid answers for frequently resolved domain names.

For the curious reader: Many recent versions of Windows, and Unix-link operating systems such as Linux and FreeBSD have a tool called *nslookup*. Often installed as a command line application, nslookup allows you to manually perform DNS queries and explore your local network's DNS configuration. Similar tools may be found under names like *dig* or *host*.

References

[ISO3166] http://www.iso.org/iso/en/prods-services/iso3166ma/02iso-3166-code-lists/list-en1-semic.txt.

[RFC768] J. Postel, Ed, "User Datagram Protocol", RFC 768. August 1980.

[RFC791] J. Postel, Ed, "Internet Protocol Darpa Internet Program Protocol Specification", RFC 791. September 1981.

[RFC793] J. Postel, Ed, "Transmission Control Protocol", RFC 793. September 1981.
[RFC826] D.C. Plummer, "An Ethernet Address Resolution Protocol", RFC 826. November 1982.
[RFC1112] S. Deering, "Host Extensions for IP Multicasting", RFC 1112. August 1989.
[RFC1191] J. Mogul, S. Deering, "Path MTU Discovery", RFC 1191. November 1990.
[RFC1390] D. Katz, "Transmission of IP and ARP over FDDI Networks", RFC 1390. January 1993.
[RFC1519] V. Fuller, T. Li, J. Yu, K. Varadhan, "Classless Inter-Domain Routing (CIDR): an Address Assignment and Aggregation Strategy", RFC 1519. September 1993.
[RFC1591] J. Postel, "Domain Name System Structure and Delegation", RFC 1591. March 1994.
[RFC1771] Y. Rekhter, T. Li, "A Border Gateway Protocol 4 (BGP-4)", RFC 1771. March 1995.
[RFC1918] Y. Rekhter, B. Moskowitz, D. Karrenberg, J. de Groot, E. Lear, "Address Allocation for Private Internets", RFC 1918. February 1996.
[RFC1981] J. McCann, S. Deering, J. Mogul, "Path MTU Discovery for IP version 6", RFC 1981. August 1996.
[RFC2003] C. Perkins, "IP Encapsulation within IP", RFC 2003. October 1996.
[RFC2004] C. Perkins, "Minimal Encapsulation within IP", RFC 2004. October 1996.
[RFC2050] K. Hubbard, M. Kosters, D. Conrad, D. Karrenberg, J. Postel, "Internet Registry IP Allocation Guidelines", RFC 2050. November 1996.
[RFC2131] R. Droms, "Dynamic Host Configuration Protocol", RFC 2131. March 1997.
[RFC2132] S. Alexander, R. Droms, "DHCP Options and BOOTP Vendor Extensions", RFC 2132. March 1997.
[RFC2225] M. Laubach and J. Halpern, "Classical IP and ARP over ATM", Internet Request for Comment 2225. April 1998.
[RFC2328] J. Moy, "OSPF Version 2", RFC 2328. April 1998.
[RFC2453] G. Malkin, "RIP Version 2", RFC 2453. November 1998.
[RFC2473] A. Conta, S. Deering, "Generic Packet Tunneling in IPv6 Specification", RFC 2473, December 1998.
[RFC2616] R. Fielding, J. Gettys, J. Mogul, H. Frystyk, L. Masinter, P. Leach, T. Berners-Lee, "Hypertext Transfer Protocol – HTTP/1.1", RFC 2616. June 1999.
[RFC2821] J. Klensin, Ed, "Simple Mail Transfer Protocol", RFC 2821. April 2001.
[RFC2993] T. Hain, "Architectural Implications of NAT", RFC 2993. November 2000.
[RFC3022] P. Srisuresh, K. Egevang, "Traditional IP Network Address Translator (Traditional NAT)", RFC 3022. January 2001.
[IANAP] Internet Assigned Numbers Authority, "Directory of General Assigned Numbers (last viewed January 2006)", http://www.iana.org/numbers.html.
[8023] IEEE Std 802.3. "IEEE Standards for Local and Metropolitan Area Networks: Specific Requirements. Part 3: Carrier Sense Multiple Access with Collision Detection (CSMA/CD) Access Method and Physical Layer Specifications", 1998.

5

Network Latency, Jitter and Loss

Regardless of game genre, the realism of online game play depends on how well the underlying network allows game participants to communicate in a timely and predictable manner. In the previous chapter, we broadly reviewed the nature of modern IP network services. In this chapter we will discuss in more detail three characteristics of best effort Internet Protocol (IP) service – latency, jitter and loss – that have a significant impact on game play experience and game design. We will also look briefly at the technical methods Internet Service Providers (ISPs) can utilise to control these characteristics of their network services.

5.1 The Relevance of Latency, Jitter and Loss

As noted in the previous chapter, IP packets carry information between sources and destinations on the network. *Latency* refers to the time it takes for a packet of data to be transported from its source to its destination. In many networking texts, you will also see the term *Round Trip Time* (RTT) in reference to the latency of a round trip from source to destination and then back to source again. In many cases the RTT is twice the latency, although this is not universally true (some network paths exhibit asymmetric latencies, with higher latencies in one direction than the other). In online gaming communities, the term *lag* is often colloquially used to mean RTT.

Variation in latency from one packet to the next is referred to as *jitter*. There are a number of mathematically precise ways to define jitter, usually depending on the timescale over which the latency variation occurs and the direction in which it occurs. For example, a path showing an average 100 ms latency might exhibit latencies of 90 ms and 110 ms for every alternate packet – fairly noticeable jitter in the short term, even though the long-term average latency is constant. Alternatively, the path might exhibit latency, that is, drifting – 90 ms, 95 ms, 100 ms, 105 ms, 110 ms, 105 ms, 100 ms... and so on. For our purposes, it is sufficient to know that latency can fluctuate slowly or rapidly from one packet to the next.

Figure 5.1 summarises the key difference between latency and jitter.

Packet loss refers to (not surprisingly) the case when a packet simply never reaches its destination. It is lost somewhere in the network. A path's packet loss characteristic is often described in terms of packet loss rate or packet loss probability (ratio of the number of packets lost per number of packets sent).

Networking and Online Games: Understanding and Engineering Multiplayer Internet Games
Grenville Armitage, Mark Claypool, Philip Branch

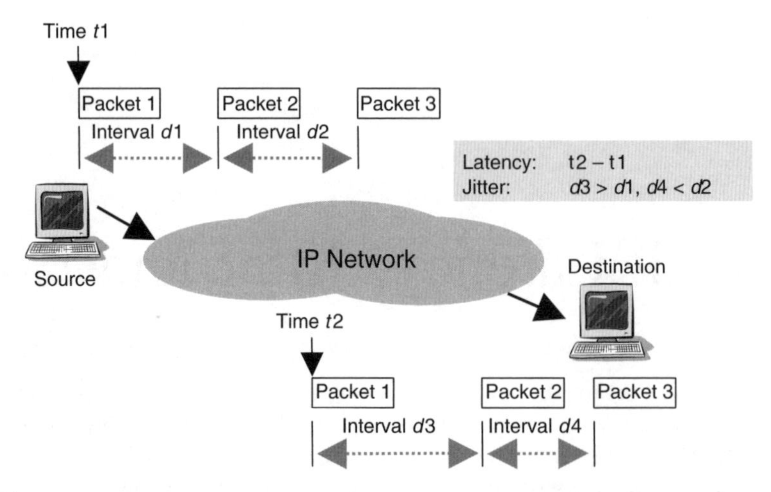

Figure 5.1 Latency and jitter affect streams of packets travelling across the network

All three have a negative impact on online game play. Latency affects the absolute sense of real-time interactivity that can be achieved within the game context. The latency of a network path between client and game server puts a lower bound on how quickly game-state information can be exchanged and consequently limits each player's ability to react to situational changes within the simulated game world. Jitter can make it difficult for players (and the game engine itself) to compensate for long-term average latency from the network. Jitter must be kept as low as possible. The consequences of packet loss should be self-evident – game-state updates are lost, and the game engine (at client and/or server) must cover up the loss as best as it can.

In the rest of this chapter, we will review the various sources of latency, jitter and loss inside ISP networks, and briefly outline the technical methods ISPs can utilise to reduce and control these characteristics.

5.2 Sources of Latency, Jitter and Loss in the Network

Three main sources of delay add cumulatively to the total latency experienced by a packet.

- Finite propagation delays over large distances (we must obey the laws of physics)
- Serialisation delays (particularly over low bit rate links)
- Congestion-related queuing delays.

A number of mechanisms introduce jitter by causing variations in latency from one packet to the next.

- Path length changes
- Packet size variations and
- Transient congestion.

Packet losses are typically due to:

- Excess transient congestion causing queues to overflow;

- Link layer bit errors causing packet corruption or
- Routing transients temporarily disrupting the path.

5.2.1 Propagation Delay and the Laws of Physics

The speed of light dictates the top speed with which any information can propagate in a particular medium (air or optical fibre, for example). The speed of transmission along copper wires and cables is usually less than the speed of light in air, depending on the particular physical construction of the cables (for example, the dielectric properties of the insulation). Since the speed of light is finite, the laws of physics impose a lower bound on latency between geographically distant points on the Internet. We refer to this latency as *propagation delay*.

As the speed of light is roughly 299,792 kilometres per second, propagation delays become noticeable over links spanning many thousands of kilometres or where the path hops through a number of routers each thousands of kilometres apart. For example, a 12,000-km path (roughly Sydney to Los Angeles in an airplane) would exhibit at least 40-ms latency (or 80-ms RTT) simply because of the finite speed of light. Most game players will come across this issue when they are connected to servers in different states or countries. It is also possible to find a high latency network path between two geographically close sites if the sites connect to the Internet via different ISPs (as noted in Chapter 4).

A rough rule of thumb for propagation delay is

```
latency (ms) = (distance of link in kilometres)/300
```

(If the speed of light in the medium is less than 300,000 kilometres per second the latency will be higher. This would be the case, for example, in optical fibre where the speed of light is about 30 % slower than that in a vacuum.)

5.2.2 Serialisation

Serialisation occurs in many real-life situations. Crowds of people getting on a bus go through the door one at a time; we board planes one at a time; a worker loads crates onto a truck one at a time; and the one remaining bank teller who has not taken a lunch break can only process us one at a time. Serialisation occurs on most link layers, and is another source of latency in IP networks.

Most link technologies are, at their lowest level, serial in nature. Frames are broken into sequences of bytes, and the bytes are sent one bit at a time. The finite period taken to transmit an IP packet one bit at a time is referred to as *serialisation latency*. This period of time depends on the speed of the link (in bits per second) and the length of the packet being sent. Serialisation latency adds to any speed of light delays experienced by a packet.

Depending on the link layer technology, there might be extra bits at the beginning and end of each byte (traditional serial ports, for example) or at the beginning and end of each frame (standard Ethernet LANs, for example). Thus, the total serialisation latency experienced by an IP packet also depends on the framing protocol used by a particular link layer. Consider the time taken to transmit a 1500-byte IP packet on a 100-Mbps Fast

Ethernet LAN and a nominally '56-Kbps' V.90 dial-up connection using Point-to-Point Protocol (PPP) [RFC1661]):

- On the Ethernet link, a 1500-byte IP packet becomes 1526 bytes long (8 bytes of ethernet preamble, 12 bytes for source and destination MAC address, two bytes ethernet protocol type and 4 bytes trailing Cyclic Redundancy Check [CRC]), or 12,208 bits. At 100 Mbps, it takes 122 microseconds to transmit the frame containing this packet.
- On a V.90 dial-up link, the uplink is limited to 33.6 Kbps while the downlink rarely exceeds 51 Kbps. If we further assume PPP encapsulation of 8 bytes, the 1500-byte IP packet requires $1508 \times 8 = 12{,}064$ bits to transmit. Thus, a 1500-byte IP packet takes 359 ms to transmit towards the ISP (upstream) and 237 ms towards the client (downstream).

Serialisation latency is primarily an issue with low-speed links common in consumer access networks (for example, dial-up modem service or consumer Asymmetric Digital Subscriber Link, ADSL). Using the above numbers, even a small 40-byte IP packet (48 bytes including PPP overhead) takes 11.4 ms on a V.90 upstream and 7.5 ms on a V.90 downstream (contributing 19 ms to any RTT measured using 40-byte packets).

A similar situation occurs on high speed links when your ISP imposes temporary rate caps. For example, consider an ISP using ADSL2 + to offer 4 Mbps downstream service and a customer who has exceeded the download limit for the month. The ISP temporarily applies a rate of 64 Kbps until the end of the month, imposed at the IP packet level. Although each packet is still individually transmitted at 4 Mbps, the ISP achieves a 64-Kbps long-term rate cap by limiting the number of packets per second that can be sent. The effective serialisation delay is as though the ADSL2 + link was literally running at 64 Kbps.

Serialisation latency should only be calculated once (at one end of the link) since the receiving end is pulling bits off the link at the same rate that the transmitting end is sending them. Aside from a slight offset in time due to propagation delay, the transmission and reception processes occur essentially concurrently.

A rough rule of thumb for serialisation delay is

```
latency (ms) = 8*(link layer frame length in bytes)/(link
   speed in Kbps)
```

(Note that for some link technologies, such as Asynchronous Transfer Mode (ATM) and Data over Cable Service Interface Specification (DOCSIS), the relationship between link layer frame length and IP packet length is nonlinear and nontrivial to calculate.)

5.2.3 Queuing Delays

One of the core underlying assumptions of the Internet's best effort philosophy is that everyone's traffic is largely bursty and uncorrelated, allowing us to benefit from a concept known as *statistical multiplexing*. Multiplexing occurs when multiple inbound streams of packet traffic converges on a single outbound link at a particular router or switch. The inbound packets are multiplexed (interleaved in time) onto the outbound link.

However, unlike traditional telephone company networks IP routers do not prearrange guaranteed timeslots on the outbound link for the competing inbound packet streams.

Statistical multiplexing assumes that everything will be okay if the average bit rate of all the inbound packet streams does not exceed the capacity of the outbound link. Or alternatively stated, 'most of the time most packets do not arrive at the same time and will not collide with each other's need for the outbound link'. Of course, in reality, packet arrivals do coincide. When multiple inbound packets arrive at the same instant for the same outbound link, the packets are queued up and transmitted one after the other. We will refer to this situation as *transient congestion.*

As previously noted, transmitting a single packet on a physical link introduces a finite serialisation delay proportional to the link's speed. Consequently, any packet queued up for transmission on a particular link will experience additional latency due to the serialisation delays of every packet in the queue ahead of it. We refer to this as *queuing delay*.

Queuing delays appear under many guises in everyday life. Teller service at your local bank, or check-in at your favourite airline, involve queues to cope with customer arrival patterns that are bursty and that often exceed the processing capacity of the available tellers or check-in agents. The delay you personally experience can be short or long, depending on how many people arrived just before you and how fast the tellers (or check-in agents) are processing previous customers.

In a typical consumer environment, queuing delays are seen when multiple computers on a home LAN try to send packets out through the same cable modem or ADSL modem. When outbound packets converge on the broadband router they will be queued up, waiting their turn to be transmitted on the upstream link to the ISP (which is usually ten to a hundred times slower than the local LAN link).

Another form of queuing delay occurs on shared links where only one host can transmit at a time, and a link access protocol operates to share transmission opportunities amongst attached hosts. A modern example involves 802.11 b/g wireless LANs (so-called "*WiFi*" networks). The Carrier Sense Multiple Access/Collision Avoidance (CSMA/CA) mechanism and four-way handshake protocol (to avoid hidden-node problems) create access variable delays that depend on the traffic load on the wireless network (number of clients and/or number of packets per second being sent). For example, 802.11 b networks have been shown experimentally to add 50 to 100 ms to the RTT when heavily loaded by bulk TCP file transfers [NGUYEN04].

5.2.4 Sources of Jitter in the Network

As noted in Chapter 4, the actual path taken by a stream of packets can vary over time. When a route change occurs, the new path may be shorter or longer (in both kilometres and number of hops). Packets sent immediately after the route change will still get to their destination and yet experience a different latency. Route changes are usually uncommon, but can create a noticeable change in lag between a game client and server.

On links that introduce noticeable serialisation delay, we can experience jitter due simply to the variations in size between consecutive packets sent over the link. This relates directly to congestion-induced queuing delay. Queuing delay depends entirely on the statistical properties of other traffic sharing a congested outbound link – not just when and how fast the competing packets arrive, but their size distribution too. Because transient congestion depends on the vagaries and burstiness of entirely unrelated traffic, the queuing delay seen by any particular flow of packets can seem entirely random.

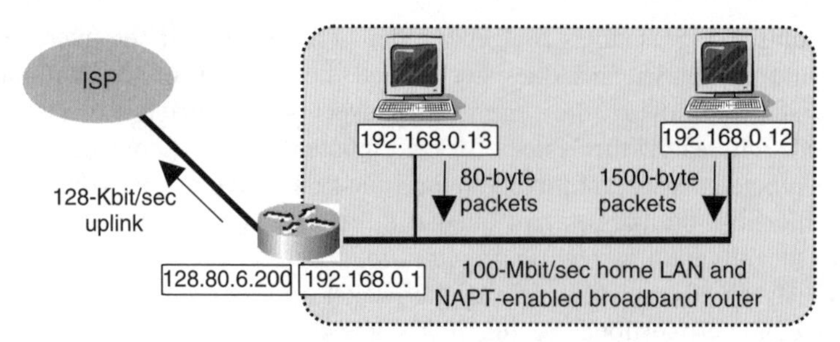

Figure 5.2 A congested uplink can introduce jitter through queuing streams of different sized packets

Consider the case of two home computers on a single 100-Mbps Ethernet LAN, communicating to the outside world over a 'broadband' 128 Kbps link (perhaps early cable modem or ADSL service, Figure 5.2).

Host 1 is generating a stream of 80-byte IP packets, one every 40 ms. Assume (for the sake of argument) link layer overheads on the broadband link add a fixed 10 bytes, making the frame 90 bytes long. These frames take 5.6 ms to transmit at 128 Kbps. Now host 2 suddenly decides to transmit a random stream of 1500 byte IP packets, with a mean interval of 500 ms. These larger packets arrive at the cable or ADSL modem and take roughly 94.4 ms to transmit. From host 1's perspective, its stream of 80-byte IP packets now experience random jitter – much of the time the packets go through immediately, but every so often there are a couple of packets that are delayed by up to 94.4 ms in excess of their usual transmission time. When host 2's 1500-byte packet arrives at the broadband modem, subsequent packets from host 1 (which are arriving every 40 ms) must be queued for up to 94.4 ms while the 1500-byte packet is transmitted. After that, the queued 80-byte packets are transmitted (in 94.4 ms at least two packets are likely to have arrived from host 1). Host 1's packets who were queued suffer additional latency relative to their siblings who arrived while the broadband modem's queue was empty.

There is one more source of jitter that only affects IP over PPP/High-level Data Link Control (HDLC) [RFC1662]) over serial links. HDLC framing implements a technique known as *byte stuffing* to ensure reliable identification of frame boundaries over serial links. In simple terms, whenever the byte 0x7E appears in the IP packet, it is replaced by two bytes 0x7D-0x5E for transmission over the serial link. Thus, for example, a User Data Protocol (UDP)payload containing only the value 0x7E repeated 200 times would appear to be a 238-byte UDP/IP/PPP frame. However, the two hundred 0x7E bytes would each be doubled, resulting in a 438-byte frame being transmitted over the link. The associated serialisation delay would be that of a 438-byte frame rather than a 238-byte frame. In short, the serialisation delay over such links can be randomly influenced by the unknowable distribution of HDLC control bytes in the IP packets.

In Chapter 10, we will look at typical game traffic packet sizes and inter-packet intervals – information that can help estimate latency and jitter over various links.

5.2.5 Sources of Packet Loss in the Network

A packet may be lost at many different points within the network, and for a number of entirely different reasons.

First, at the physical layer all links experience a finite (albeit usually extremely low) rate of data corruption – which we refer to as *bit errors*, and characterise by a link's *bit error rate*. Bit errors may be introduced by poor signal-to-noise ratios in the digital-to-analogue-to-digital conversion process, resulting in erroneous encoding or decoding of data. Bit errors may simply be due to electrical glitches in hardware. Some link technologies encode additional information within each frame to enable limited reconstruction of a frame after one- or two-bit errors. This is known as *forward error correction*, (FEC). In any case, uncorrected bit errors are usually discovered through cyclic redundancy check (CRC) calculations at both the transmitting and receiving end of a link. The CRC is a 16- or 32-bit value mathematically calculated during transmission and sent along with each frame, and then recalculated at the receiver. If the original and recalculated CRCs differ, the frame is discarded. Since this can occur anywhere along an IP path, there is no way to inform the end hosts why or how their packet was lost.

Second, transient congestion can become so severe that queuing points along the path simply run out of space to hold new packets. When this happens new packets are simply dropped until the queue(s) have emptied enough to take new packets. This is the network's most aggressive form of self-protection against too much traffic. Some networks even employ proactive packet drop mechanisms that introduce a random, nonzero loss probability well before the queue is full. (This is referred to as *active queue management*, with the most well-known variant known as *Random Early Detection* (RED) [RFC2309]). Proactive packet dropping is intended to force TCP-based applications to slow down before congestion becomes too serious, but they have little effect on non-reactive UDP-based game traffic (except to cause packet loss).

Third, dynamic routing changes do not always converge immediately on a fully functional and complete end-to-end path. When route changes occur, there can be periods of time (of tens of seconds to minutes) where no valid shortest path exists between previously connected sources and destinations. This manifests itself as unexpected packet loss affecting tens, hundreds or thousands of packets.

5.3 Network Control of Lag, Jitter and Loss

Online games, particularly the real-time interactive genres, need greater control over network latency, jitter and loss than more traditional email, online 'chat' and web surfing applications. In this section, we will briefly review the mechanisms that ISPs can deploy to control network conditions on behalf of game players, and the difficulties faced by ISPs in utilising these techniques effectively.

One approach is for ISPs to ensure that their link and router capacities far exceed the offered traffic loads, and to utilise creative routing of traffic to ensure that no single router becomes a point of serious congestion. Attractive because of its conceptual simplicity, this approach tends to be practical only for large or core network operators who have flexible access to physical link infrastructures (for example, a telephone company that owns both an ISP and the underlying optical fibre between cities).

The 'just deploy more bandwidth' approach tends not to work where high speed technologies simply cannot be deployed in a cost-effective manner (for example, where 100-Mbps home LANs meet sub-1 Mbps 'broadband' access links, as in Figure 5.2). ISPs and consumers must contemplate ways of prioritising access rather than simply hoping for the best. This means identifying some IP packets as worthy of 'better' service than other IP packets, and this carries with it many questions of how to differentiate between types of IP packets and what constitutes 'better' service.

5.3.1 Preferential IP Layer Queuing and Scheduling

There is not a lot we can do about packet loss due to bit errors except invest in and install high-quality link technologies. On the other hand, we can provide some control over congestion-induced loss, latency, and jitter by preferentially handling and forwarding important game packets over and above other traffic.

Preferential treatment occurs in many real-world situations. Airlines have separate check-in counters for business and first-class passengers. Freeways have high-occupancy or peak-period lanes restricted to cars carrying multiple passengers. Police and emergency services vehicles can gain temporary preferential treatment and road access by turning on their sirens and warning lights.

Central to any of these schemes is the goal of letting some people 'jump the queue' and get ahead of others. This requires three steps – classification (to identify who or what deserves preferential treatment), separate queuing (to isolate those getting preferential treatment) and scheduling (to actually provide the preferential treatment). The same applies in IP networks. Congested routers need to classify, separately queue and preferentially schedule some packets over others. Output ports that previously used only one queue will now have two (or more) queues – one for normal best effort service, and another for high-priority packets.

Many edge and access network routers today are capable of classifying IP packets using five pieces of information from inside each packet:

- The packet's source and destination IP addresses,
- The packet's protocol type, and
- Source and destination port numbers (if the packet is TCP or UDP).

This is sometimes referred to as *flow classification, microflow classification* or *5-tuple classification.* A set of rules in each router dictates which combinations of IP address and port numbers are considered priority packets and which are not. On the basis of these rules, every packet can be classified into a high-priority or normal priority queue. Ultimately, the scheduler decides when to transmit packets from each queue.

Serialisation delays still apply to each packet transmission, and congestion-induced transmission delays and the potential for packet loss still exist. However, these events now occur on a per-queue basis. By splitting traffic into separate queues, we isolate the game traffic from much of the queuing and serialisation delays that afflict regular best effort traffic.

Consider the two cases in Figure 5.3. Five packets converge on a congested router interface – the first four are just regular traffic, the fifth packet is associated with an active game. In the absence of preferential queuing and scheduling, all five packets will be queued and transmitted in order or arrival. However, imagine if the router implements two queues – normal and high priority. The first four packets can be placed in the normal traffic queue. The fifth packet is recognised as a game packet, placed in the high-priority queue, and is transmitted as soon as the current packet from the normal queue finishes

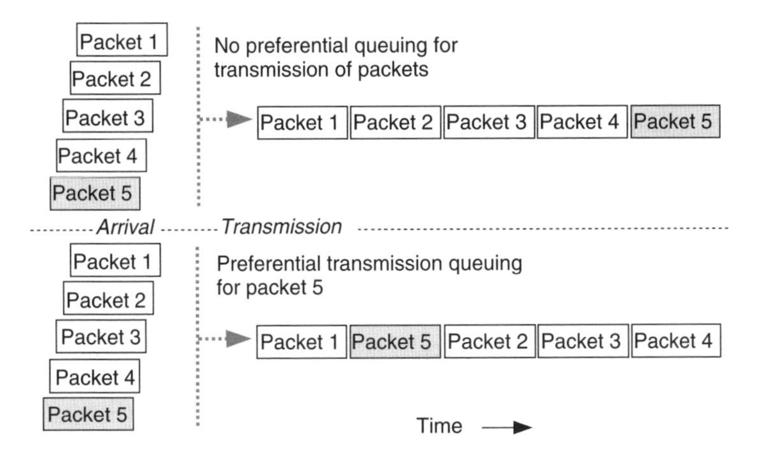

Figure 5.3 Preferential queuing and scheduling can allow priority packets to 'jump the queue'

transmission. The game packet has effectively jumped the queue of normal packets. (A real-world analogue would be a police car going around a queue of cars and trucks waiting to enter a single-lane toll road.)

Because many scenarios involve only two priority levels, routers may also choose a simpler classification scheme based on a 6-bit Differentiated Services Code Point (DSCP) embedded in the IP header [RFC2474, RFC2475]. Often the 5-tuple classification occurs near the edges of an ISP network, and the result is encoded into the packet's DSCP field. Core routers then use simplified classification rules based on the DSCP to assign packets into priority and normal queues.

Packet loss is also controllable on a per-queue basis. Each queue can be allocated different amount of space, ensuring that game packets can still be queued even when the normal traffic queue is full and starts to drop packets. Each queue may implement a different form of active queue management [RFC2309].

We have deliberately simplified this discussion to focus on the key elements. There are many different variations on classification, queuing and scheduling algorithms, which we do not need to explore here. Interested readers might refer to a book on IP quality of service for more details (for example [ARM2000]).

5.3.2 Link Layer Support for Packet Prioritisation

Unfortunately, classification, queuing and scheduling at the IP packet layer does not solve all our problems when facing serialisation delays on low-speed links. Consider the real-world case of a police car jumping a queue of cars and trucks for entry into a single-lane toll road. The 'normal traffic' is a mixture of short cars and long articulated trucks. Yet no matter how much priority we give the police car, if a long truck commenced onto the toll road an instant before the police car arrived, it may still have to wait awhile at the entrance to the single-lane toll road. There is no way for the police car to pull back the truck, or slice the truck in half, in order to avoid waiting.

The same occurs when, for example, an ADSL modem only prioritises upstream traffic into game and normal categories at the IP packet level. In Figure 5.3, if packet 1 was

1500 bytes long and had begun transmission just before packet 5 arrived, no amount of preferential queuing at the packet level could avoid packet 5 having to wait until packet 1 is transmitted.

A partial solution to this scenario is offered by link layer technologies capable of concurrently interleaving multiple IP packets during transmission.

For example, most consumer ADSL services actually run ATM over the ADSL physical layer, and then implement their packet service over a single ATM virtual channel [TR017, ARM2000]. However, in principle, the ADSL modem could utilise two or more parallel virtual channels upstream towards the ISP, and send different streams of cells on each virtual channel [I311]. Game packets and normal packets would be segmented onto distinct virtual channels and their constituent cells interleaved in time according to the packet's priority. Recall that ATM cells are only 53 bytes long (carrying 48 bytes of payload and 5 bytes of header). Thus, the serialisation delay experienced by a game packet drops down to the time it would take to send an ATM cell belonging to the normal packet already in flight. It is as though the police car could cut the truck in half.

Unfortunately, ATM over ADSL is not used in this manner as of the time of writing, although it remains a distinct possibility as ISPs grapple with offering the best control of jitter and latency on the upstream side of consumer broadband links. Similar approaches are theoretically possible with Multi-Class Multi-Link (MCML) and Real-Time Framing (RTF) PPP over serial links [RFC2686, RFC2687] and segmenting packets across minislots in DOCSIS-based cable modem systems [DOCSIS]. Further discussion of such schemes is beyond the scope of this book.

5.3.3 Where to Place and Trust Traffic Classification

Much of the classification, queuing and scheduling techniques discussed so far are practical with today's technology. The major stumbling block faced by ISPs today actually revolves around how they know, for certain, what packets to give priority to at any given time. (Or, in the terminology of this chapter, which packets are game traffic deserving preferential treatment.) Conversely, how does the ISP ensure that only authorised traffic gets preferential treatment?

Only a game client and game server know for sure what IP packets constitute game traffic. The ISP wishing to provide preferential service must configure their routers with the 5-tuple rules or DSCP values that identify packets deserving preferential treatment. But how does the ISP initially discover the right 5-tuple or DSCP values? Ask the game clients for 5-tuple values? Trust the game clients to set the DSCP to a 'well-known' value?

In general, ISPs should not trust an outside entity (which includes hosts it does not control) to set the DSCP value appropriately. If it becomes well known that 'DSCP == 1' results in priority traffic handling, every application on every host will naturally be inclined to set DSCP to one on their outbound packets. We would be right back at the beginning of the problem, with everyone's packets classified into the game traffic queue.

Thus the assignment of DSCP values must be performed by routers the ISP trusts, typically at the edges of the ISP's network. DSCP assignment will occur after 5-tuple classification by the ISP's router. This begs an obvious question – how does the ISP's router know which 5-tuples represent game traffic?

If an ISP hosts its own game server, then it knows one side of the 5-tuple – the IP address and port number of the game server. This information may well be sufficient to perform classification at the edges and assign trusted DSCP values to game packets as they progress across the ISP's network. (Any packet heading to or from the game server's IP address and port is considered a game packet regardless of the client's IP address and port number.)

In principle, the game client might use a signalling protocol to inform the ISP, on a case-by-case basis, when a new flow of game packets begins. This would inform the ISP of the specific 5-tuple associated with any particular game traffic. However, at the time of writing this book, no ISP implemented any such IP signalling protocol that is of use to a game client. (And if they did, there would be a number of difficult questions regarding authentication of signalling messages to ensure that only legitimate game clients were telling the ISP 'this 5-tuple represents game traffic'.)

An emerging approach is for ISPs to automatically detect game traffic by looking for particular statistical properties rather than specific 5-tuple values or well-known port numbers. Once a flow has been identified as having the statistical properties of game traffic, the flow's 5-tuple can be passed out to routers along the path who need to provide preferential treatment [STE2005].

When the ISP classifies on a 5-tuple, it becomes difficult for an opportunistic client to create packets that "look like" game traffic even though they are not. Given that the source and/or destination IP addresses are a key part of the 5-tuple classification rules, the opportunistic packets would have to actually be going to or from a known game server in order to gain preferential treatment. This is unlikely to be generally useful to non-game applications.

Given that there are hundreds of ports on which game traffic might exist, and thousands of IP addresses that might represent game servers, ISPs and game developers face an interesting challenge to correctly, safely and securely identify game traffic for preferential treatment. This is very much an open question, unsolved as of the time of writing.

5.4 Measuring Network Conditions

We can use the 'ping' command to get an approximate sense of the RTT between one of our hosts and one of the other endpoints on the Internet. Ping is available on UNIX-derived systems (for example, FreeBSD and Linux) and Microsoft Windows systems. Ping uses Internet Control Message Protocol (ICMP) packets [RFC792] to probe a specified remote host and to measure the time it takes for the probe packet to go out and return. There is a caveat regarding ping – many routers handle ICMP messages differently from regular packets, and thus ping can sometimes return RTT estimates that are higher (by a few milliseconds) than the RTT that would be experienced by actual TCP or UDP traffic along the same path. Nevertheless, running multiple repeated pings against a fixed, remote IP host can reveal paths where the RTT fluctuates around a reasonably consistent average value. Ping defaults to using a small, 64-byte IP packet. However, the user can force the use of larger ICMP packets in order to reveal the possible effect of serialisation delay along the path.

Another tool that can provide insight into network path characteristics is 'traceroute' (under UNIX-derived systems) or 'tracert' (under Windows). Traceroute attempts to estimate the RTT to each router hop from your host to a nominated destination host. The difference in RTT reported at different numbers of hops away from your host can reveal links that include significant propagation delays. Traceroute can assess paths using different sized probe packets, which can be used to reveal serialisation delay sensitivities along a path. (As of writing, Windows XP's 'tracert' did not allow configuration of the probe packet's size.)

Both traceroute and tracert can resolve the IP addresses into domain names of routers seen along a path. If the ISPs have used human-readable names, you can sometimes infer things about the geographical path being followed (e.g. if city names are encoded in the domain names of the router interfaces seen by traceroute).

There are many web sites around the Internet that offer traceroute facilities from their location. You can find a list of pointers to such sites at http://www.traceroute.org

For example, Figure 5.4 shows one of the listed sites (located at Telstra in Canberra, Australia) revealing the effect of propagation delay when performing a traceroute to www.lucent.com (192.11.226.2, based in New Jersey, USA). The traceroute output shows about 5 ms between first and seventh hop, then ~145 ms between seventh and eighth hops (from Sydney to Los Angeles) followed by another jump between Los Angeles and New York of ~70 ms (essentially next door to New Jersey).

Note that the ~145-ms RTT between Sydney and Los Angeles is substantially higher than the ~80 ms estimate in Section 5.2.1. This can be attributed to the lower propagation speed of signals in optical fibre, queuing delays over multiple hops and the fibre's indirect physical path between Sydney and Los Angeles being much longer than 12,000 km.

Another approach is to use information gathered by a modified game server. Figure 5.5 shows one published experiment where RTT samples from an active Quake III Arena server were used to plot the average jitter (per map played) against the average latency (per map played) [ARM2004]. One of the paper's conclusions was that jitter typically never exceeded 20 % of the average latency. Another observation from Figure 5.5 is that there are two broad clusters – one where jitter is low regardless of latency (out to 300 ms latency) and another where jitter is roughly proportional to latency. The former is attributed to paths where most latency is propagation delay. The latter is attributed to

```
 1  FastEthernet6-0.civ-service1.Canberra.telstra.net
    (203.50.1.65)  0.225 ms  0.193 ms  0.268 ms
         [..]
 7  i-7-0.syd-core02.net.reach.com (202.84.221.90)  5.457 ms
    5.636 ms  5.349 ms
 8  i-0-0.wil-core02.net.reach.com (202.84.144.101)  153.923 ms
    153.935 ms  154.057 ms
          [..]
11  0.so-3-0-0.CL1.LAX15.ALTER.NET (152.63.117.90)  153.84 ms
   154.67 ms  154.303 ms
12  0.so-5-0-0.XL1.NYC9.ALTER.NET (152.63.0.174)  227.725 ms
   265.947 ms  227.653 ms
         [..]
17  192.11.226.2 (192.11.226.2)  228.585 ms  229.29 ms  227.779 ms
```

Figure 5.4 Traceroute from Australia to the USA showing long-haul propagation delays

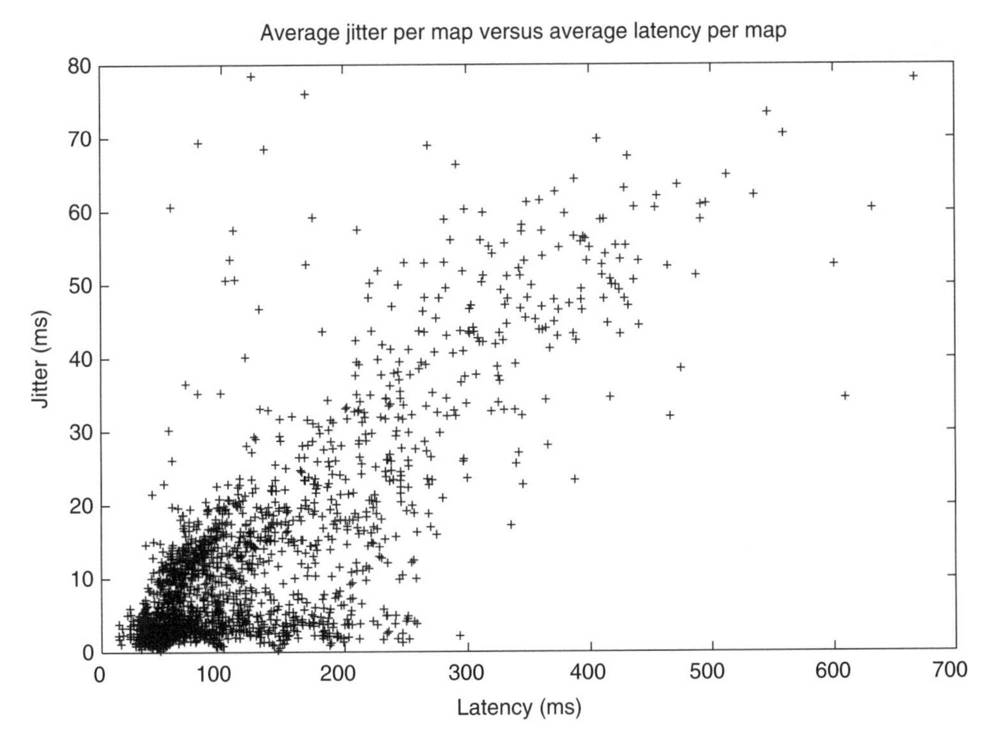

Figure 5.5 Jitter versus latency measured from an active Quake III Arena server in Australia [ARM2004]

paths where delay accumulates from many router and link hops – also places more likely to contribute to jitter.

References

[ARM2000] G. Armitage, "*Quality of Service in IP Networks: Foundations for a Multi-Service Internet*", Macmillan Technical Publishing, April 2000.

[ARM2004] G. Armitage, L. Stewart, "Limitations of using Real-World, Public Servers to Estimate Jitter Tolerance of First Person Shooter Games", *ACM SIGCHI ACE2004 Conference*, Singapore, June 2004.

[DOCSIS] CableLabs, "Data-Over-Cable Service Interface Specifications Radio Frequency Interface Specification SP-RFIv1.1-I01-990311", 1999.

[I311] ITU-T Recommendation I.311, "B-ISDN General Network Aspects", August 1996.

[NGUYEN04] T.T.T. Nguyen, G. Armitage, "Quantitative Assessment of IP Service Quality in 802.11b Networks and DOCSIS networks", *Australian Telecommunications Networks & Applications Conference 2004*, (ATNAC2004), Sydney, Australia, December 8–10, 2004.

[RFC792] J. Postel, "Internet Control Message Protocol", STD 0005, RFC 792. September 1981.

[RFC1661] W. Simpson Ed, "The Point-to-Point Protocol (PPP)", STD 51, RFC 1661. July 1994.

[RFC1662] W. Simpson Ed, "PPP in HDLC-like Framing", STD 51, RFC 1662, July 1994.

[RFC2474] K. Nichols, S. Blake, F. Baker, D. Black. "Definition of the Differentiated Services Field (DS Field) in the IPv4 and IPv6 Headers." RFC 2474. December 1998.

[RFC2475] S. Blake, D. Black, M. Carlson, E. Davies, Z. Wang, W. Weiss. "An Architecture for Differentiated Services." RFC 2475. December 1998.

[RFC2309] B. Braden, D. Clark, J. Crowcroft, B. Davie, S. Deering, D. Estrin, S. Floyd, V. Jacobson, G. Minshall, C. Partridge, L. Peterson, K. Ramakrishnan, S. Shenker, J. Wroclawski, L. Zhang, "Recommendations on Queue Management and Congestion Avoidance in the Internet", RFC 2309. April 1998.
[RFC2686] C. Bormann, "The Multi-Class Extension to Multi-Link PPP", RFC 2686. September 1999.
[RFC2687] C. Bormann, "PPP in a Real-time Oriented HDLC-like Framing", RFC 2687. September 1999.
[STE2005] L. Stewart, G. Armitage, P. Branch, S. Zander, "An Architecture for Automated Network Control of QoS over Consumer Broadband Links", *IEEE TENCON 05*, Melbourne, Australia, 21–24, November, 2005.
[TR017] ADSL Forum, "TR-017: ATM over ADSL Recommendation", 1999.

6

Latency Compensation Techniques

The delay (and variance in delay) illustrated in Chapter 5 comes from both delay caused by processing updates and delay caused by the network. While processing delay can be reduced by hardware and game algorithm improvements, networking delay is harder to reduce.

> 'There is an old network saying: 'Bandwidth problems can be cured with money. Latency problems are harder because the speed of light is fixed – you can't bribe God.'
>
> —David Clark, MIT

The first section of this chapter presents the *need* for latency compensation techniques against the backdrop of a basic client–server game architecture. Next, we talk about how the architecture can be enhanced using *prediction*, both for the server response to user input and for the behaviour of other units, to mitigate the impact of latency. Manipulation of game time using *time delay* and *time warp* can equalise gameplay for players with disparate latencies to the server. Various *data compression* techniques primarily reduce game bitrates, but in doing so can also reduce latency by reducing the time to process network data. Although not network related, *visual tricks* can reduce the user awareness of any remaining residual latency. Lastly, latency compensation can also have an impact on *cheating* and *cheat detection*, and so extra precautions must be taken.

6.1 The Need for Latency Compensation

While there has been substantial work in network congestion control toward reducing queuing delay at routers and reducing packet loss (thus avoiding extra delays added by any retransmissions), substantial delays remain, especially for network connections over the Internet to "last-mile" hops in home residences. Some see broadband as a solution for the latency problems with online gaming. However, while broadband generally offers significant improvements in network bitrates, there are other factors in network designs that limit the benefits of broadband.

(a) Despite the promise of higher bitrates, some broadband solutions still have periodic high latencies or significant packet loss during congestion.

Networking and Online Games: Understanding and Engineering Multiplayer Internet Games
Grenville Armitage, Mark Claypool, Philip Branch

(b) It will be quite some time before broadband is globally adopted by all online game players.
(c) There is an increasing number of mobile, wide-area wireless environments that cannot benefit from the fixed infrastructure of broadband to the home.
(d) Despite technology improvements, a certain amount of network delay will always remain because of the speed of light in fibre. For example, coast-to-coast delays across the United States of America cannot be less than about 25 milliseconds, not counting any additional delays from buffering. Intercontinental times will be even higher.

Most games today basically run on a client–server architecture with a single, authoritative server that handles the game logic. The clients collect input from the user and send it to the server. The server computes game state, sending the revised game state information back to the client. The client then renders the new game state to the player, and the process repeats. In some cases, the server can also act like a client, allowing a player to use the same machine as the server. In this case, the player offers to 'host' a game. That player's computer then acts like a server for all players, as well as a client for the current player. This is still fundamentally a client–server architecture, even though it may look like client–client (or peer-to-peer) on the surface, in that all machines are being used as game clients.

To illustrate the effects of latency on this basic architecture, consider the client to be where the user input is taking place. In the basic client–server model, the client sends a message to the server when user input is received. The server processes (and validates) the input and sends the results back to the waiting client to render on the local display. Thus, the players' actions are lagged by the round-trip latency between client and server. When the client acts in this manner, it is often called a 'Dumb' client since it only acts on commands from the server, and is depicted in Figure 6.1.

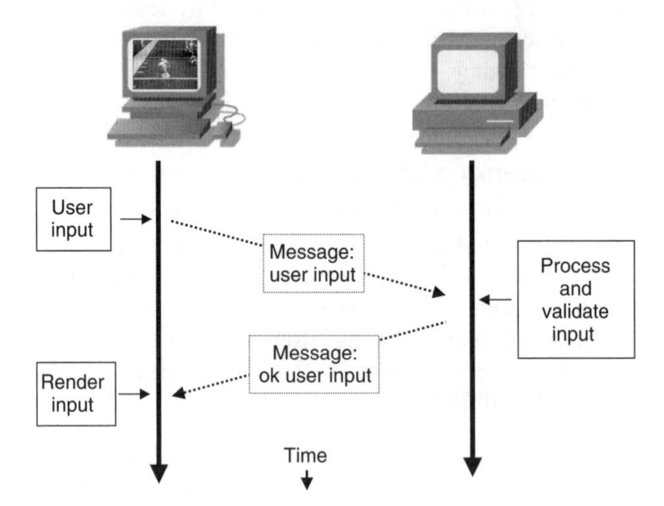

Figure 6.1 Basic client-server with 'dumb' client. Client only renders output of user input when it receives input 'ok' from server

Basic Algorithm (Client Side)

- Sample user input
- Pack up data and send to server
- Receive updates from server and unpack
- Determine visible objects and game state
- Render scene
- Repeat.

If the latency between the client and server is large enough, the user is aware of the delay between the commands given to the game and the response to the game state. As an example of this impact, consider an online American football game (e.g. Madden NFL Online) in which the player is responding to events on the screen but the input is lagged to the server by about a second. In Figure 6.2, the running back is running toward the left side of the field to avoid the defender. The user sees that there is an open lane along the sideline and pushes the controller up to run between the defender and the sideline. However, because of the latency, the processing of this input is delayed by the round-trip time to the server (one second, in this example) so that the command is actually processed after the runner goes out of bounds. Because of the latency, the user failed to gain as many yards on this running attempt as she/he would have if there was no latency.

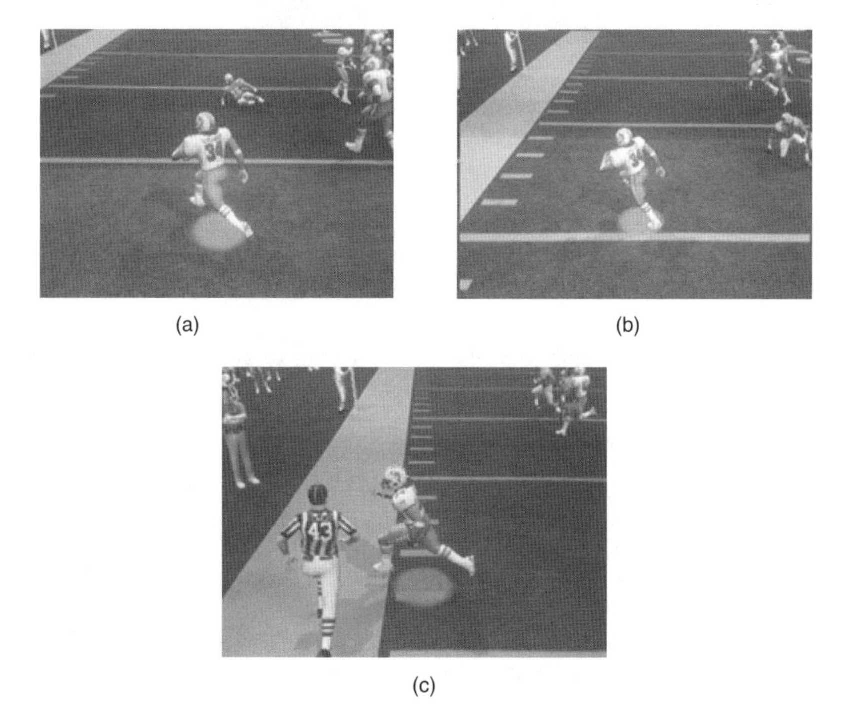

Figure 6.2 Illustration of the effects of latency on running (Madden NFL 2003). (a) User is pressing left and the player moves left. (b) User is pressing up, but player continues left because of latency. (c) Running back goes out of bounds! User curses

Figure 6.3 Illustration of the effects of latency on passing (Madden NFL 2003). (a) User is pressing throw, but throw is not processed yet because of latency. (b) Throw starts processing here because of the latency. (c) Defender intercepts throw! User curses

As another example, in Figure 6.3, the user is trying to get the quarterback to throw ('pass') the ball to a receiver. The receiver might only be away from a defender ('open') for a short window of time before crossing the boundary and being too close to the defender. At the start of play, as the receiver begins his route, the user presses the pass button in order to deliver the pass to reach the receiver at the boundary between defenders. The receiver should catch the ball at the boundary since he is open. However, because of the latency, the processing of the quarterback throwing the ball actually begins too late. By the time the ball reaches the receiver, the receiver has fully crossed the boundary and the defender catches the ball instead (an 'interception').

These examples, and many more like them in real-time game genres of all kinds, occur quite frequently in online games unless latency compensation techniques are deployed.

6.2 Prediction

Instead of waiting for the server to respond to each client action before rendering it, the client can predict the server response, allowing the game client to respond to user input immediately, rendering player and opponent movements before getting authoritative responses from the server.

Broadly, there are two categories of prediction that can take place. In the first prediction category, the client takes input from the player and predicts the server response related to only the player's units. We call this 'Player Prediction'. In the second prediction category, the client predicts the location of units that are not controlled by the player, being controlled either by other players or by a computer. We call this 'Opponent Prediction' (even if some of the other units are not necessarily on another team).[1]

6.2.1 Player Prediction

The client can predict the server response, allowing the game client to respond to user input and render player actions before getting the authoritative response from the server. This allows the game to appear immediately responsive to the user input, not needing to take a round-trip to the server and back before impacting the game. In fact, the response can be as fast as a non-networked game, thus completely removing any network latency. Using prediction, however, means the game state on the server (and the state on other client machines) will differ somewhat from the game state on the client. The amount it differs depends on the round-trip latency and, to some extent, the user actions taken. The client must therefore fix up any discrepancies in the game state when it finally does get a response from the server. The player prediction process is depicted in Figure 6.4.

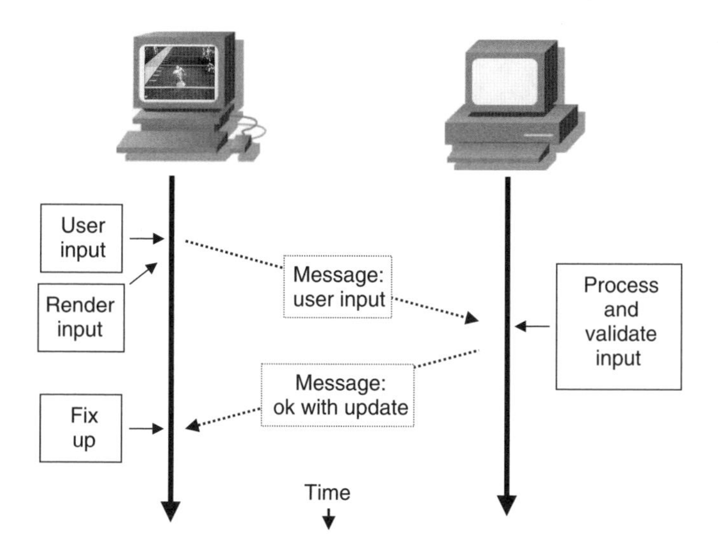

Figure 6.4 Client-server in which client renders input with predicted state before getting 'ok' from server

[1] Prediction of units controlled by others is often called 'Dead Reckoning' [Dead reckoning, CLC99, DF98], but that name does little to help remember the technique, hence our term 'Opponent Prediction'.

Predicted Algorithm

- Sample user input
- Pack up data and send to server
- Determine visible objects and game state
- Render scene
- Receive updates from server and upack
- Fix up any discrepancies
- Repeat.

As an illustration of the effects of latency on this architecture, consider the screenshots in Figure 6.5, in which both game consoles (the client and server) are connected to the same television. In the figure, the client's display is the larger picture, while the server's display is inset in the picture-in-picture. The client puts a man in motion (causing a football player to go from left to right across the screen), the result is that the client sees the in-motion player movement first, and subsequently, the player is one or two steps ahead on the client's display than he is on the server's display. We have manually drawn a box around the man in motion on each display to indicate the player of interest. Notice how the boxed player for the client in the large picture is further to the left than the boxed player for the server in the inset picture.

The notable additional step the player prediction approach has that the basic client–server approach does not is the step to 'fix up any discrepancies'. These discrepancies may occur because the server still has the master copy of the game state, and this master copy may differ from the predicted copy the client has. As an example, suppose the user moves an avatar to the right. The game at the client sends this movement

Figure 6.5 Depiction of state inconsistency (Madden NFL 2003). The large picture showing the first player's view differs from the second player's view, shown by the smaller, inset picture. A black box (drawn manually, not by the game) highlights the main difference

command to the server, a trip that may take hundreds of milliseconds. Before receiving the server's response, the client game then predicts the new location for the avatar and renders the new position on the right, thus appearing very responsive to the player input. Sometime later, when the server receives the command, it checks if the avatar was allowed to move to the right. If so, the client is allowed to move, and the new state is sent to the client where it is confirmed. If not, perhaps because another avatar had moved to the same spot first, effectively blocking the way, the server would respond to the client that the move to the right was not allowed, providing the correct game state to the client. Upon receiving the server update, the client would have to fix the discrepancy between the server's master view and the client's predicted view, ultimately rendering the correct world to the client.

The benefits of client-side prediction to responsiveness are tremendous – an online game can effectively feel as responsive as a single-player, non-networked game, short-circuiting any network latency. However, fixing up the discrepancies between the actual, server-controlled game state and the client-side predicted game state can be equally destructive – having the display of an avatar abruptly changed by rendering the correct actual scene over the incorrect predicted scene (essentially 'warping' or 'rubber-banding' the world back to the correct state from the incorrect one) can be jarring, impacting gameplay and greatly reducing immersiveness.

Fundamentally, while client-side prediction allows the game to be more responsive, it trades off consistency between the game state at the client with the game state at the server. Figure 6.6 depicts the trade-off between responsiveness to the user input and consistency to the server view of the world. On the right is the basic client–server approach, in which the user view of the world always follows the view of the world presented by the server, being consistent with the world state as computed by the server. On the left is client-side prediction, where the view of the world is predicted by the client, thus sometimes causing inconsistencies in the world state at the server and the world state at the client.

6.2.2 Opponent Prediction

With opponent prediction, the location of a unit that is controlled by another player (or computer) is estimated. The estimation starts with the last known position of the unit and computes its current, predicted position based on the speed and direction it was travelling. This predicted position is then used unless and until the unit owner sends an update of the new location, speed or direction or both. This update would be sent when the unit owner determines that the other clients cannot accurately predict the position within a predetermined threshold. The update sent contains the correct position and orientation as well as velocity vectors and other derivatives that the clients can use to initiate a new prediction. Figure 6.7 depicts the difference between the actual path of a unit and the predicted path computed by the other clients.

Figure 6.6 The trade-off between consistency and responsiveness in games with network latency

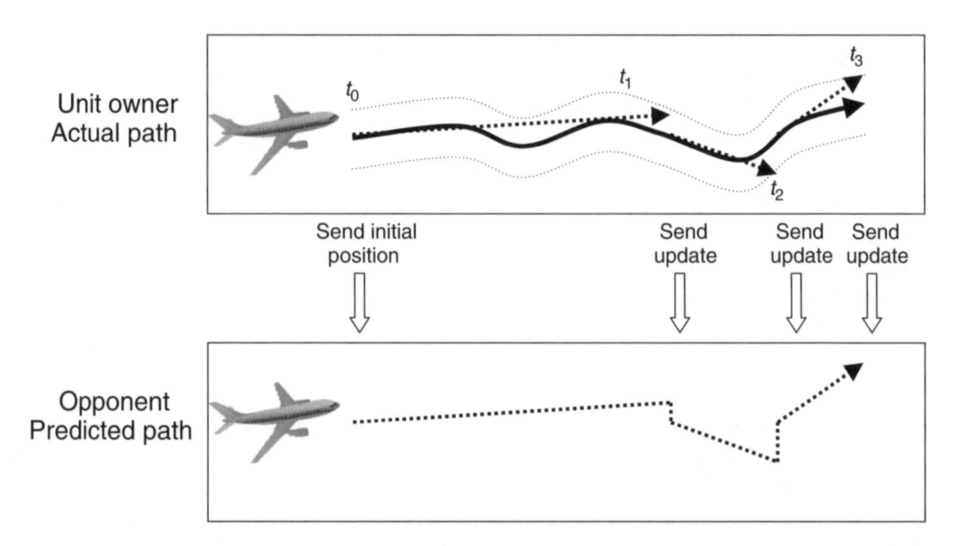

Figure 6.7 Opponent Prediction, showing actual path versus predicted path. The opponent uses the last known unit information in computing its path. The unit owner sends updates when the predicted location differs from the actual location by more than a given threshold

The top picture in Figure 6.7 shows the view by the owner of the unit (in this case, the unit is an airplane but could just as easily be a car, boat or even a human). The solid line in the middle shows the actual path as the player controlling the unit would see it. The two thinner, dashed lines that run parallel to the middle line represent the threshold for the opponent predictions. The thicker dashed lines represent the unit owner's record of the opponent's prediction. If the unit owner's predicted location of the airplane goes outside this threshold, the unit owner sends a message to all opponents with an update on the new position and heading. In Figure 6.7, after the initial position and heading is sent at time t_0, updates are sent at t_1, t_2, and t_3. The bottom picture depicts the view that opponents would see. The opponent uses the last known position and heading to predict the unit location until an update is received, whereupon the new position and heading are used.

Smaller values for the update threshold can provide more fidelity in opponent predictions at the cost of requiring more frequent updates (hence, higher bandwidth and processing overhead). Larger values for the update threshold decrease prediction fidelity, and also decrease the update rate (also decreasing the bandwidth and processing overheads). The optimal values for the threshold depend on the game, the network and processing capacities of the clients and, to some extent, the player preferences.

In making the predictions on location, there are two commonly used unit estimation algorithms from the Distributed Interaction Simulation[2] standard, both derived from basic Physics. The first predicts the location of a unit on the basis of its last known position and its velocity at that time. The second predicts the location of a unit on the basis of the last known position, velocity and acceleration. More sophisticated algorithms can use

[2] Distributed Interactive Simulation (DIS) is a protocol to support large-scale simulations developed by the Defense Advanced Research Project Agency (DARPA) Simulation Network project [Ney97, DIS].

roll, pitch, heading and even do predictions on different parts of a unit independently (e.g. the angle of a tank turret can be predicted independently of the prediction of the location of the tank itself).

Assuming $x(t)$ is the position at time t and the last update for a unit's position was received at time t_0. With the simplest form of opponent prediction, a client could assume the location of a unit at time t_1 is the same as the location of a unit at time t_0.

$$x(t_1) = x(t_0)$$

Assuming a constant velocity (v), using the velocity at time t_0, a slightly more sophisticated algorithm would predict the location of the unit to be:

$$x(t_1) = x(t_0) + v^*(t_1 - t_0)$$

Adding information about a constant acceleration (a), the location of the unit would be predicted to be:

$$x(t_1) = x(t_0) + v^*(t_1 - t_0) + (1/2)a^*(t_1 - t_0)^2$$

In general, the opponent prediction algorithm for the unit owner looks like:

- Sample user input
- Update {location | velocity | acceleration} on the basis of new input
- Compute predicted location on the basis of previous {location | velocity | acceleration}
- If (current location – predicted location) < threshold then
  - Pack up {location | velocity | acceleration} data
  - Send to each other opponent
  - Repeat.

While the opponent prediction algorithm for the opponent would look like:

- Receive new packet
- Extract state update information {location | velocity | acceleration}
- If seen unit before then
  - Update unit information
- Else
  - Add unit information to list
- For each unit in list
  - Update predicted location
- Render frame
- Repeat.

In general, units with high inertia are easy to predict (i.e. a rock rolling down a hill or a player in free-fall from an airplane), while models with little inertia are harder (i.e. a pixie with 360 degrees of movement freedom or an avatar that can teleport). Game-specific prediction algorithms can even be crafted. For example, a real-time strategy game may define what it means for a unit to 'return to base'. As long as the unit continues to return to base, all clients can accurately predict the position over time without any updates.

Opponent prediction does have its costs. It requires that each client run an algorithm to extrapolate the location of each unit for each frame rendered. If units behave unpredictably, such as is often the case in a frenetic first-person shooter game with permissive physics constraints, the benefits of trying to predict unit locations diminish.

While opponent prediction is presented here as a means to reduce latency, in many cases, it can also greatly reduce bitrates. For large-scale simulations with relatively static units, or units that move in a very predictable fashion, the use of opponent prediction can eliminate the interchange of most state update messages. Even for computer-controlled units that use random variables to add variety to their actions, as long as the clients have the same prediction algorithms and same initial random seed, their predictions can remain faithful even through "random" behaviour. And for unreliable network messages, such as over UDP, the use of opponent prediction can help smooth over lost update messages. In these cases, the trade-off is using more CPU cycles at each client to reduce the latency in updating units and, in some cases, the bitrate required in exchanging messages.

One approach of managing any inconsistent behaviour in predicted game state information when an update arrives is to have the game clients use any new state update information as part of a 'rendezvous' with the goal of achieving that new state in a short amount of time [CF05]. With this rendezvous idea, instead of immediately jumping to the new state, rules, customised for each game, determine how the game world is to be updated to the new representation. For example, if it is determined that the position of a soccer ball is incorrect, the game may move a computer-controlled player to kick the ball to get the ball to the correct location.

Another aspect of opponent prediction is the potential for unfairness in the case where players have heterogeneous latencies. In general opponent prediction schemes, an update to a position is provided when the predicted position deviates by a threshold from the actual position. If an update on the actual position is sent to several players, the players that are farther away, in terms of network latency, will get the update later, making the difference in the predicted position and the actual position larger than the difference in the same for updates sent to closer players. This results in unfairness in that closer players have more accurate game world representations than farther player.

This unfairness can be removed completely by the use of Time Delay (see below), whereby the update to closer players are held up by exactly enough to make their effective latency the same as the farthest player [AB05]. This results in an additional trade-off. Namely, minimising the unfairness of differences in the accuracy of opponent prediction among players, while maximising the amount of prediction error all players receive.

Another method of reducing unfairness is to send more frequent updates to the players that are farther, thus reducing the error in their predicted versus actual prediction [AB05]. If a budget on the total update rate is required, the update rate to the closer players can be reduced to make the update rate to the farther players higher.

6.2.3 Prediction Summary

In general, any game state predicted at the client may differ from the real, authoritative game state at the server because:

(a) users are controlling the other states and it may be difficult, or even impossible, to predict in a precise way what action the user will take. The prediction may be that another player's avatar will continue running straight, but the other player suddenly moves the avatar left, or the avatar may be predicted to be standing still when it is actually shooting. The prediction can be constrained by the physics of the gameplay (for example, a user can only run at a certain speed, cannot fly over water, etc.), but since games are designed to give users choices, the user-controlled avatar may act in an unexpected fashion, causing the client-side view of the game state to diverge.

(b) the prediction the client may use is a simplistic approximation of the game state computation at the server. This can happen for more complex predictions. For example, the server may be computing the velocity, acceleration and rotational torque of a flying projectile to precisely compute the current location and facing of the object. A client using simplistic prediction may only use the last known location and velocity in predicting the new location, resulting in a discrepancy in the predicted and actual locations. Client-side predictions that are too simplistic can generally be solved by using more CPU cycles and more complex prediction models at the client.

6.3 Time Manipulation

It must be noted that even without prediction, game states rendered at the clients will differ from each other. This is because it takes some time (about a half a round-trip time) for a client to receive the world state from the server. When one client is further away (has a higher round-trip time) from the server than another client, there may be unfairness in the game play. For example, suppose two players finally defeat a monster they had been battling. The server, controlling the now defeated monster, generates some treasure as a reward for the battle and drops it on the ground for the players to pick-up. A message with the location of the treasure is sent to the clients for each of the players. Suppose the first client is quite close (in terms of network latency) to the server, so the player sees the treasure and acts upon it immediately, moving to gather the loot. The second client is farther away from the server and thus responds more slowly, in fact, after the first player has moved to get the treasure. This unfairness can degrade the gameplay for many games.

Even though many online games are real-time, the actual progression of time in the game can be manipulated to account for disparities in latency among clients. Two techniques to do this are a *time delay* for all commands to handicap players with a low latency, and a complementary technique called *time warp* to accommodate players with a high latency.

6.3.1 Time Delay

A common technique for dealing with differences among clients is to delay processing and sending of user commands to equalise latency. Essentially, instead of processing client commands right away, the server delays them for some time, allowing a client that is further away (in terms of network latency) to respond to the game state. In essence, this allows both the clients to have the same effective latency in providing updates to the game world at server with a delay as large as the buffer chosen by the server. World-state updates sent by the server can also be delayed in being sent out, sending the update to the client that is further away before sending updates to the client that is closer. Or, the

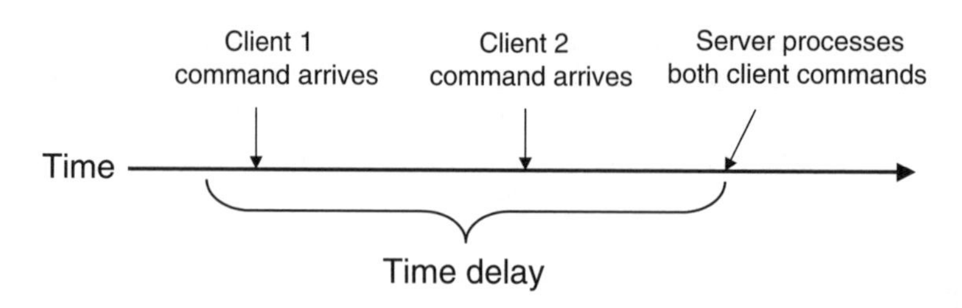

Figure 6.8 Time Delay. A game server can hold client commands for a fixed amount of time, the *time delay*, before processing in order to compensate for clients with higher latency than others

clients themselves can use buffering to equalise the latency, with the client that is closer to the server delaying processing of the server state updates, while the client that is further away acting on the updates immediately upon receiving them.

Thus, buffering can provide fairness between clients with disparate latencies, but at the cost of making the gameplay less responsive. A related technique at the client is called *interpolation*, where a client has rendered the local world, but has a server update showing the position of units at a later time. Instead of immediately rendering the latest server world state, the client can interpolate the world at intermediate states, allowing the local world state to progress smoothly to the server world state. Another technique commonly used with interpolation is *extrapolation* where the world state at a future time is predicted. Extrapolation is, in fact, another form of prediction presented above.

In choosing a time delay buffer, it should be noted that users often prefer a consistent, even if large, delay, rather than a variable delay that jumps about widely. The effects of variable delays in terms of response time for interactive processes (e.g. Telnet or Internet phone) is well-known. In fact, Internet phone applications make use of time delay to explicitly avoid varying delays. Thus, the size of any time delay chosen should be adjusted infrequently (on the order of tens of seconds), even if the latency measurements from the server change more frequently.

6.3.2 Time Warp

A successful, widely used time manipulation mechanism is to have a server to rollback (or *time warp*) the events in a game to the time when a client command was inputed [Mau00, Valve]. The player provides input on the basis of the current state of the game at the client. Because of the lag between the client input and the server receiving the command, the state at the server may have changed. For example, the player shoots at an opponent at time t_0, but by the time the message arrives at the server at time t_2, the opponent had moved at time t_1. Using time warp, the server rolls back the events it had processed since the client provided the input (roll back to time t_0 in the above example). In this case, the server might determine that this older event has a bearing on subsequent events, changing their effect to make the global world state consistent. For example, the server may determine the player had hit and killed the opponent, meaning the opponent movement at time t_2 was invalid.

The general algorithm for the server is as follows:

- Receive packet from client
- Extract information (user input)
- Elapsed time = current time – latency to client
- Rollback all events in reverse order to current time – elapsed time
- Execute user command
- Repeat all events in order, updating any clients affected
- Repeat.

Note that for time warp to be effective, it requires an accurate measurement of the latency between a client and server in order that the game time can be rolled back the proper amount. Fortunately, the frequent message exchanges between server and client provide many opportunities for client–server latencies to be refined as the game is played.

By using time warp, clients that have a high latency to the server can still have their commands executed in the correct game-time order without impacting other players through a time delay. This allows players to respond to the current state of the game world without having to account for latencies to the server. For example, with time warp, a player in a first-person shooter can aim directly at an opponent, not having to worry about the opponent moving before the server gets the shot update message. Without time warp, in the same first-person shooter a player would have to 'lead' an opponent by aiming in front of them in order to hit the opponent when they did move.

The popular first person shooter game, Half-Life 2 (HL2), makes use of time warp [Valve] (along with some other techniques mentioned in this chapter). For testing purposes, the HL2 server allows additional lag to be added to the clients. The server administrator can observe the actual location of a unit and the location for the unit with time warp by having a separate client on the same machine as the server (a listen server). Figure 6.9 shows a screenshot of an HL2 listen server. The round-trip latency to the client is 200 ms, meaning the user's commands are executed 100 ms before the screenshot. The

Figure 6.9 Example of Time Warp (Half-Life 2). The target is on the left and is ahead (in terms of time) of the client. The boxes on the right represent the targets the client had when shooting (back in time) and that the server uses when time is warped back to determine a hit

grey boxes show the target position on the client where it was 100 ms ago. Since then, the target moved to the left while the user's command was travelling over the network to the server. When the user command arrives at the server, the server rolls back time (time warp) to put the target in the position it was at the time the user shot, indicated by the black boxes. The server determines there was a hit (the client sees blood from the wounds). Note that the client (grey) and server (black) boxes do not match precisely because of small differences in the time measurements and the speed of the moving target.

Time warp can cause some inconsistencies, however. Suppose a player places the cross-hairs of a gun on an opponent and fires. The server, using time warp, will ultimately determine this is a "hit". However, in the meantime, because of client–server latency, the opponent may have moved, perhaps even around a corner and out of sight. When the server warps time back to when the shot was fired and determines the opponent was shot, it will seem to the opponent that the bullets actually bent around the corner. Fortunately, this disconcerting effect is minimised if the opponent cannot see the attacker or if the opponent was still in the open and not hiding.

6.3.3 Data compression

While the propagation delay between a client and a server may be fixed (ultimately bounded by the speed of light), reducing the size of the messages sent between client and server can also reduce latency. As noted in Chapter 5, a small packet has a shorter transmission time than a large packet because of serialisation delay, a component of delay that accumulates at each router between the server and the client.

There are several ways that packet sizes can be reduced.

(a) *Lossless Compression.* Data can be manually compressed using well-known data algorithms. Unlike compression techniques for audio, video and images, compression for game data must be lossless, meaning all the bits that are compressed must be restored when uncompressed. Lossless data compression finds repeated patterns in the bits and compresses the repeats to use fewer bits. Most algorithms used for lossless compression are based on techniques developed by Lempel and Ziv in 1977 and later refined by Terry Welch's 1984, and are hence called 'LZW algorithms' [Wel84].

(b) *Opponent Prediction.* As mentioned above, for some game units, opponent prediction can greatly reduce the amount of data that needs to be transmitted between game clients.

(c) *Delta Compression.* Rather than sending complete state information whenever there is a change to the world or its units, it may be possible to send updates as changes (or *deltas*) from the previous world. This technique requires reliable delivery of data (such as, by using TCP as the transport protocol), but can be effective when the entire game world update is large but the changes are small.

(d) *Interest Management.* Instead of sending all data to each client, only a subset of data, the data that is of interest to the client, can be sent [BGRP01, MBD00]. The area of interest for a client is called the *aura* and is where the interaction between the client and other game units occurs (Figure 6.10). Auras need not be symmetric, where the *focus* of one object needs to intersect the *nimbus* of another object in defining the aura. The goal of interest management is to reduce the number of messages needed to be sent to every client.

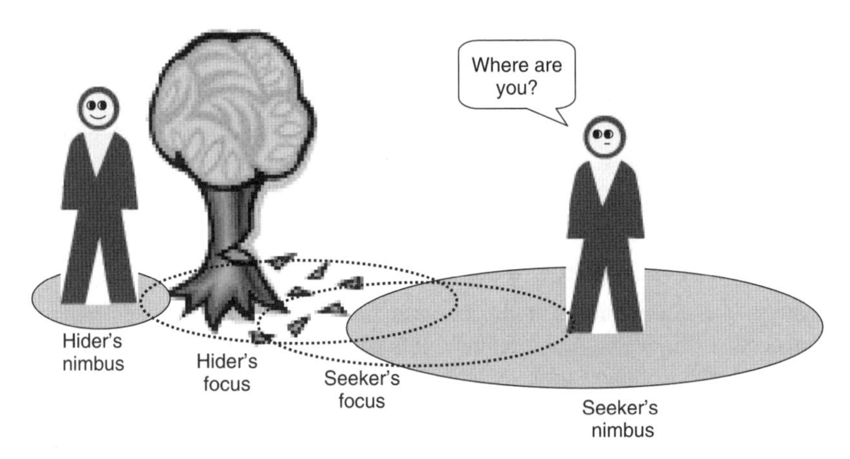

Figure 6.10 Aura of interest, illustrated by the game 'Hide and Seek.' The Aura is made up of a Focus and Nimbus. If the Focus of a unit intersects the Nimbus of another unit, they can interact. Here, the Hider can see the Seeker, but the Seeker cannot see the Hider

(e) *Peer-to-Peer.* Using a peer-to-peer network where clients send data directly to each other rather than to the server can reduce bitrates to the server. Peer-to-peer architectures are used for some common game aspects, such as voice chat during a game, but can be extended to include all non-essential game aspects (e.g. players could customise their avatar appearance, sending these appearances only changes directly to other peers).

(f) *Update Aggregation.* Sending updates after some periodic delay can avoid some of the network overhead associated with each message. For example, if player A moves at time t_0 and player B moves at time t_1, rather than send two messages to player C, the server may choose to send one, slightly larger message at time t_1 containing the moves for A and B, thus avoiding the network overhead for packing and sending an additional message. In some cases, update aggregation can even reduce the number of messages sent. For example, if A turns an avatar to face left and then turns to face right, the server may choose to elide the commands and send only the latest facing to the right to other clients.

In general, using compression trades off CPU cycles at the client and server for reduced network load. If the network reduction is significant, and the compression and decompression is not computationally prohibitive, compression makes sense as a latency compensation technique.

6.4 Visual Tricks

This section mentions a few techniques that do not really involve networking, but can cover up network latency to the user.

A start-up animation can be used to hide latency from the client to the server. For example, if a boat gets ready to move, the game may require it to visually raise sails before it starts to actually move. Such animation delays can take a couple of hundred milliseconds, even allowing a message to go the server and back before the unit actually moves. This way, if the server indicates the move is not allowed (perhaps another boat is blocking the way), there is no discrepancy to fix, while the player still receives immediate feedback making the game feel responsive.

Similarly, local confirmation can be used immediately even if the remote effect is not confirmed by the server. For example, if a player pulls the trigger on a gun, the game client can play a shooting sound effect and show a puff of smoke, even if the impact of the shot is not determined for some time.

6.5 Latency Compensation and Cheating

Unfortunately, the anonymity that the Internet provides often promotes behaviour that would not occur offline. For online games, this often means the propensity for cheating. Some of the latency compensation techniques, while attractive for helping gameplay by reducing the effects of Internet latency, provide new opportunities for cheating.

It may occur to the astute reader that using opponent prediction could eliminate the need for a server completely. Each client could compute their own actions and send the needed updates to the other clients when there is a change. In fact, this model of client-side communication only is the one proposed in DARPA's DIS protocol and works quite well *if the clients can be trusted.* However, for many game conditions, especially online, multiplayer games where players do not know one another, pure-client control of the game world state is prone to cheating. For example, a client could just send a message to another client saying 'my player is right behind your player and just shot you in the head'. Without an authoritative server to confirm or refute such client actions, some games would suffer from rampant cheating.

With time warp, a client could interfere with measurements about round-trip time, making the server believe the client is further away than it actually is. This would allow the client to respond to events that, in essence, happened in the past and hence giving unfair advantage. Similarly, with time delay, a client with slow reflexes could claim a higher latency than it actually has, causing a large time delay at the server, thus neutralising the better reflexes that opponents may have. Opportunity to cheat with these time manipulations are exacerbated if the client controls the timing measurements and time delays.

Interest management, while reducing network bitrates, can also be abused by cheaters. Clients can claim interest in game state that they could otherwise not see, using this information to gain a tactical advantage. For example, in a strategy game, a client could freely scan the other side of the map, locating the opponents base and attacking when the base should be concealed by the 'fog of war'. Or, in a shooter game, a player could observe an opponent on the other side of the wall, when the opponent should be concealed.

References

[AB05] S. Aggarwal and H. Banavar, "Fairness in Dead-Reckoning Based Distributed Multi-Player Games", In Proceedings of the 4th ACM Network and System Support for Games (NetGames), Hawthorne, NY, USA, October 2005.

[BGRP01] S. Benford, C. Greenhalgh, T. Rodden, and J. Pycock, "Collaborative virtual environments", *Communications of the ACM*, Vol. 44, No. 7, pp. 79–85, 2001.

[CF05] A. Chandler, J. Finney, "On the Effects of Loose Causal Consistency in Mobile Multiplayer Games", In Proceedings of the 4th ACM Network and System Support for Games (NetGames), Hawthorne, NY, USA, October 2005.

[CLC99] W. Cai, F. Lee, L. Chen, "An Auto-Adaptive Dead Reckoning Algorithm for Distributed Interactive Simulation", In Proceedings of the 13th ACM Workshop on Parallel and Distributed Simulation, Atlanta, Georgia, USA, Pages: 82–89, 1999.

[Dead reckoning] Wikipedia, "Dead Reckoning", [Online] http://en.wikipedia.org/wiki/Dead_reckoning, Accessed 2006.

[DF98] C. Durbach and J.M. Fourneau, "Performance Evaluation of a Dead Reckoning Mechanism", Proceedings of the Second International Workshop on Distributed Interactive Simulation and Real-Time Applications, page 23, July 19–20, 1998.

[DIS] Wikipedia, "Distributed Interactive Simulation", [Online] http://en.wikipedia.org/wiki/Distributed_Interactive_Simulation, Accessed 2006.

[Mau00] M. Mauve, "How to Keep a Dead Man from Shooting", Proceedings of the 7th International Workshop on Interactive Distributed Multimedia Systems and Telecommunication Services (IDMS), 2000, pages 199–204, Enschede, Netherlands, October 2000.

[MBD00] K.L. Morse, L. Bic, and M. Dillencourt, "Interest Management in Large-Scale Virtual Environments", *Presence*, Vol. 9, No. 1, pp. 52–68, 2000.

[Ney97] D.L. Neyland, "Virtual Combat: A Guide to Distributed Interactive Simulation", *Stackpole Books*, Mechanicsburg, PA, 1997.

[Wel84] T.A. Welch, "A Technique for High Performance Data Compression", *IEEE Computer*, Vol. 17, No. 6, pp. 8–19, 1984.

[Valve] Valve Developer Community, "Source Multiplayer Networking", [Online] http://developer.valvesoftware.com/wiki/Source_Multiplayer_Networking, Accessed 2006.

7

Playability versus Network Conditions and Cheats

Ultimately, a game hosting company, an Internet Service Provider (ISP) and a game manufacturer are aiming for the same thing – satisfied consumers. Satisfaction is achieved by understanding, and avoiding, the circumstances that would undermine an enjoyable game-play experience.

In this chapter we look at methods people have used to measure and infer player tolerance for network issues (such as latency, loss and jitter). We also look at how network communication models impact on cheats and cheat-mitigation techniques available to game developers.

7.1 Measuring Player Tolerance for Network Disruptions

Quantifying the typical player reaction to network-level characteristics is a nontrivial task. Everyone *knows* that 'latency is bad for gaming'. The task for ISPs and game hosting companies is to determine just *how much* latency becomes noticeably 'bad' for some definitions of bad, and to develop similar insights for loss and jitter as well.

A player's sense of satisfaction with any particular game will usually depend on a range of environmental factors. Players may be more or less judgemental of network-induced game-play disruptions if they are tired, hungry or have pre-existing social relationships with other players on the server. Any technique for inferring a relationship between player satisfaction and network conditions must consider these other influences.

There are two distinct approaches for discovering player tolerance to network disruptions.

- Build a controlled lab environment in which to test small groups of players under selected conditions
- Monitor player behaviour on public servers over many thousands of games.

Controlled usability trials are preferable whenever possible. One can monitor (and later account for) tiredness, hunger and social relationships between players. Arbitrary and repeatable network-level latency, loss and jitter between the players and the game server are introduced artificially. By varying the network conditions and keeping other

Networking and Online Games: Understanding and Engineering Multiplayer Internet Games
Grenville Armitage, Mark Claypool, Philip Branch

environmental conditions steady, we can draw fairly solid conclusions about player tolerances from modestly small groups of players.

Unfortunately, it is often hard to find a set of people willing to sit and play in a controlled lab environment, let alone actually obtain the resources to create the lab, in the first place. The alternative is to correlate user behaviour on an existing game server with changes in network conditions over time. This approach is less than ideal, because we cannot control (or even know) the environmental factors affecting every player who joins our server. We cannot control the precise network conditions affecting each and every player. At best we can wave our hands around and invoke the 'law of large numbers' – make measurements over thousands of games, correlate player success with known network conditions and hope the remaining unknown factors cancel themselves out.

Defining a metric for player satisfaction can also be problematic. In controlled trials, every player can be asked to fill out a survey form after each game, allowing self-reporting of subjective 'satisfaction'. When running a public game server it is far harder to entice players into filling out your survey. Unless the game's developer integrates a survey mechanism into the game itself, you are asking players to make a special, separate visit to your web site after playing on your server [OLI2003]. In all cases, care must be taken to provide questions that discover a player's environmental circumstances and avoid questions that bias player's answers.

Another approach is to track objective in-game measures of success, for example, the 'frag' (kill) rate of players on a First Person Shooter (FPS) game. This can usually be done equally well with both controlled lab trials and public servers, since the information will be contained in the server's own log files.

7.1.1 Empirical Research

Over the past few years, a number of empirical trials have been run using public servers – here we will touch on work done with games such as Quake III Arena, Half-Life, Unreal Tournament 2003 and Warcraft III.

In 2001, two Quake III Arena servers were established with identical map cycles, identically limited to six players at a time, had two 'bots' at all times, and virtually identical 'server names' visible to the public [ARM2001, ARM2003]. They differed only in that one was located in Palo Alto, California (USA) and the other was located at University College London, London (UK). The Californian server ran from May 17 to August 18, 2001 and saw 5290 unique clients who accumulated a total of 164 'days played'. Of the 5290 clients, 338 clients each accumulated more than 2 hours total playing time during this period. The London server ran from May 29 to September 12, 2001 and saw 4232 unique clients who accumulated a total of 77 'days played'. Of the 4232 clients, 131 clients each accumulated more than 2 hours total playing time during this period.

Figure 7.1 illustrates the impact of a player's median latency (ping) on their average 'frag rate'. Since games run for many minutes, a fractional improvement on your frag rate can make quite a difference in your ranking relative to other players.

Another perspective on latency is provided by Figure 7.2. This shows, as a cumulative histogram, the number of games played by clients who experienced a particular median latency throughout their game. A somewhat tenuous argument can be made that players come back more frequently if the latency is tolerable, hence the drop off from

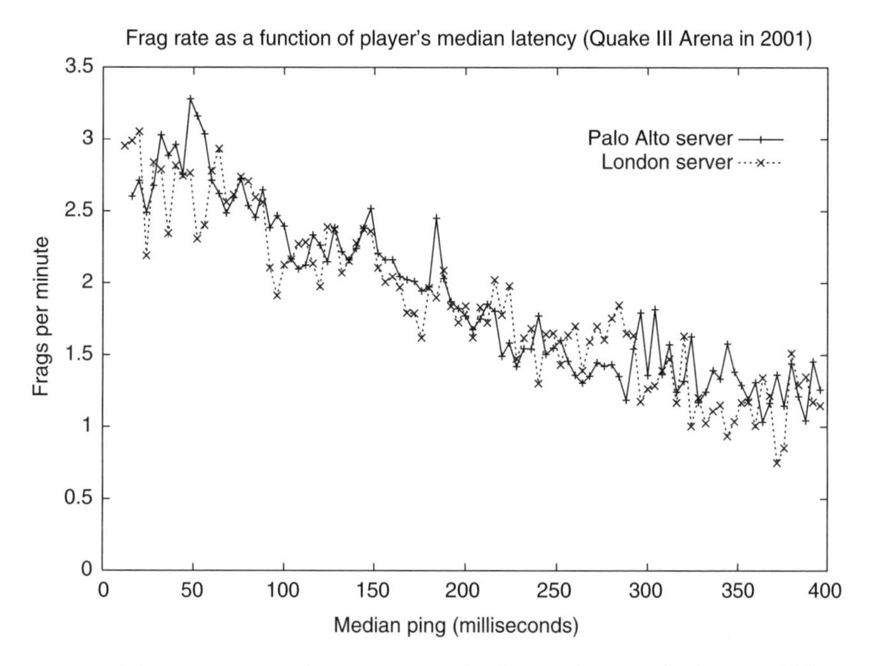

Figure 7.1 Frag rate as a function of median latency (Quake III Arena in 2001)

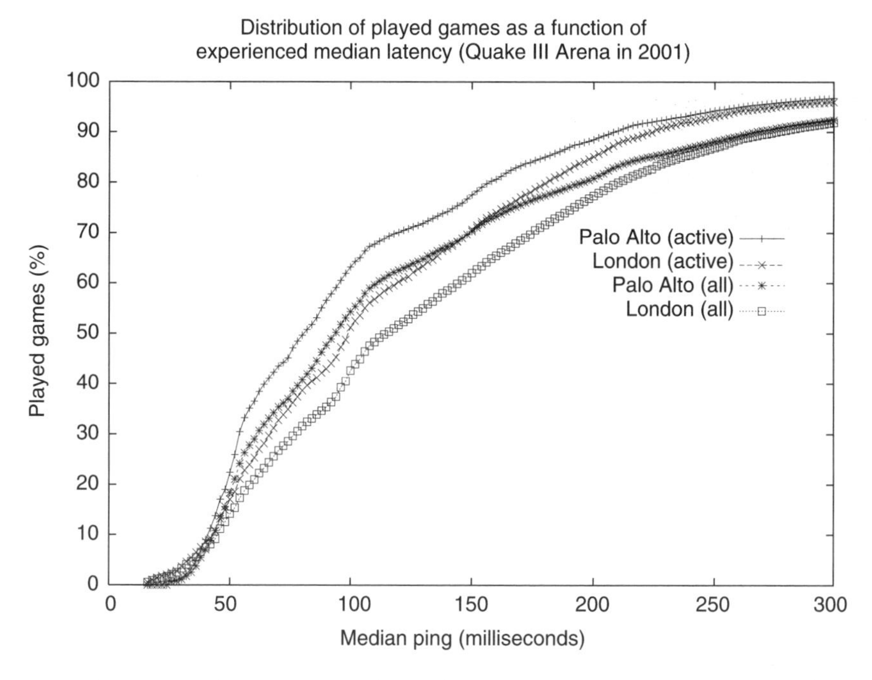

Figure 7.2 Percentage of 'played games' as a function of median latency (Quake III Arena in 2001)

games played by people, which have over 150–180 ms latency, reflects an upper bound of tolerance for players of Quake III Arena.

Unfortunately, the 2001 Quake III Arena trials lacked any information on player motivations and personal environmental conditions. They also did not measure the network jitter and packet loss rates being experienced by players who experienced particular median latencies. Thus Figure 7.1 and Figure 7.2 are more illustrative of what we might expect, and should not be taken as hard data.

Another study in 2001 was based on a public Half-Life server in the UK [HEN2001]. For over three and a half weeks the maps rotated every 60 minutes, and they saw 31,941 instances of a player joining and then leaving the server, attributable to 16,969 unique players. This study concluded that latencies around 225–250 ms dissuade players from joining a server, but below that level there was no strong evidence of absolute latency alone being a predictor of a player's likelihood of joining, staying or leaving a server. Another key conclusion was that simplistic correlation of data measured from public servers is fraught with potential errors.

Early research results led to the development of more sophisticated testbeds. In a 2003 follow-up to [HEN2001], two identical (with respect to configuration and location) public Half-Life servers were augmented with a Linux-based network switch. Using the Linux iptables and libipq functionality [IPTABLES], this switch could add controlled latency to particular client traffic arriving over the Internet [HEN2003]. Player behaviour over weeks and months was monitored as latency was alternately added to one server and then the other. When one server had 50 ms added there was a definite player-preference towards joining the server with lower latency at any given time. Interestingly, when latency was added to an individual player after they had started playing, the likelihood that the player would subsequently leave seemed to decrease the longer that player had been playing.

In another example, the latency and packet loss tolerance of 'Unreal Tournament 2003' (UT 2003, an online multiplayer FPS game) was publicised in 2004 [BEI2004]. The authors established a controlled network environment of four client hosts and a local UT 2003 server (somewhat like Figure 7.3). A Linux-based network router running NISTnet [NISTNET] was used for introducing configurable latency and packet loss to each client. Results suggested that players were unaware of packet loss rates of up to 5 %, and had only limited latency sensitivity when performing simple or complex movements. However, a distinct latency sensitivity emerged when using weapons, and players using more precise weapons were more impacted by increases in latency. [The paper's authors concluded '*... even modest (75–100 ms) amounts of latency decreases accuracy and number of kills by up to 50 % over common Internet latency ranges.*' and noted that '*latencies as low as 100 ms were noticeable and latencies around 200 ms were annoying*'.]

Of course, FPS is not the only game genre of interest. In 2003, a study was published on the latency tolerance of players in Warcraft III – a multiplayer, online Real-Time Strategy (RTS) game [SHE2003]. A controlled network environment (rather like Figure 7.3) was established, with two local Warcraft III clients and a local Warcraft III server. Latency was varied over multiple games while monitoring player behaviour during distinct 'build', 'explore' and 'combat' phases within the game. The authors concluded that '*... overall user performance is not significantly affected by Internet latencies ranging from hundreds of milliseconds to several seconds.*' In addition, Warcraft III uses TCP rather than UDP for its underlying transport protocol, ensuring that packet loss is translated into additional

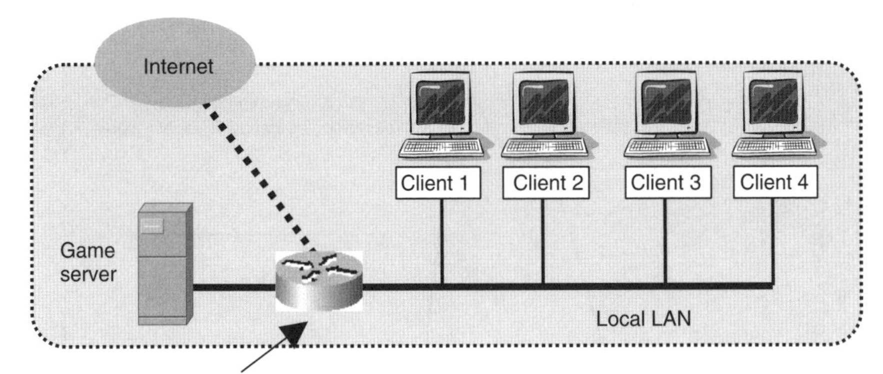

Figure 7.3 A small testbed for controlled latency and packet loss trials

latency at the application level. The evidence supports a belief that RTS games are significantly more tolerant of network latency than FPS games – success in the game depends more on strategic decisions than sheer reflex and reaction time.

Sports games have also been evaluated for latency tolerance. In 2004, a study was published on network latency and EA Sports' 'Madden NFL® Football' [NIC2004, MADDEN]. The authors used a small, dedicated lab network (along the lines of Figure 7.3) with two play stations, two consoles and a router running NISTnet for controlled delays. A player's success at 'running' and 'passing' were not noticeably affected until latency reached approximately 500 ms. The authors conclude that this game has latency requirements falling somewhere between the strict demands of typical FPS games and the looser demands of RTS games.

7.1.2 Sources of Error and Uncertainty

A number of limitations plague most published research to date in this area. Public-server studies usually lack any real knowledge about every player's external environmental conditions, and experience a limited range of network conditions. Lab-based studies usually only control network conditions and often collect an insufficient number of data points to be statistically rigorous.

Rather than being insurmountable problems, these are simply issues to be considered in future work. It is important to test player behaviour under a broad mix of network and environmental conditions. One key limitation of public servers is that you do not get much choice in the range of latency, jitter and loss rates experienced by the players. Some public-server studies have seen network-induced jitter to be almost always less than 20 % of the latency [ARM2004A]. This limits one's ability to draw conclusions about player behaviour under low-latency/high-jitter conditions because such conditions rarely occur on public servers.

Public-server trials may also suffer from variability in the RTT estimation algorithms used within each game server. The simplest way to track client latency is to modify each server to log the server's own per-client, dynamically updated RTT estimates.

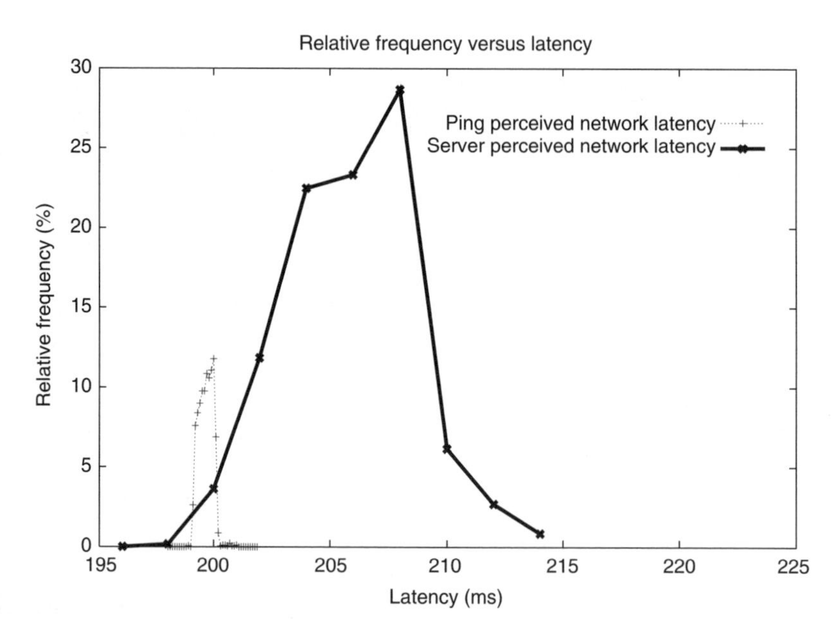

Figure 7.4 Measuring RTT with 'ping' versus a Quake III Arena server's internal mechanism

Unfortunately, the game server's RTT estimation occurs in the application space rather than kernel space, and is not necessarily strictly designed for accuracy.

Figure 7.4 shows an example where a Quake III Arena server was separated from a client by a router providing configurable network delay [ARM2004A]. Regular 'pings' (ICMP echo/reply) were run concurrently over the same link. The ICMP echo/reply technique reveals path RTT tightly sitting between 199–200 ms. In contrast, the Quake III Arena server's internal RTT estimates range between 198–214 ms (with most between 200–210 ms) – the median RTT reported by the game server is 5 ms higher than the network path's RTT, and suffers from some self-inflicted jitter.

Creating statistically relevant lab-based studies can also be rather problematic. It is not uncommon for researchers to believe that they, and a handful of close friends or colleagues, are representative of all players. Unfortunately, the rules of statistics say that many tens, if not hundreds, of independent repeat trials are required before one can draw solid conclusions.

For example, consider a generic study of player tolerance for latency and loss. A game is played with a fixed latency and loss, the player experiences are recorded, and then a new game is played again with a different latency and loss. If the estimated standard deviation of measurements for any given latency and loss settings was one (1) and we wanted to achieve 95 % confidence intervals with an error of 0.25, we would need at least 62 *independent* samples [ZAN2004]. For each and every latency and loss setting, that is a lot of samples. Particularly when you realize that, strictly speaking, 'independent samples' means using different players, not simply the same player repeating the game a number of times.

Practical considerations come into play when aiming for statistically valid lab-based studies. Consider the logistics of setting up 62 PCs with 62 licensed copies of an FPS game client and getting 62 people to give up their afternoon to play the same map repeatedly. One game for every permutation of controlled network conditions. For example, six possible latencies and three possible loss rates give you 18 repetitions of the same map. Play each map for a realistic 10–15 minutes. Now keep the 62 volunteers focused, alert, fed and somehow independent throughout the entire period.

It is hardly surprising that most published lab-based studies of FPS games have yet to come close to this level of statistical accuracy.

Finally, one might be tempted to consider a game's own lag-compensation mechanisms to be a source of uncertainty, because it hides the consequences of latency. Some games allow players to turn off lag-compensation on a game-by-game basis. However, if the game is normally shipped with lag-compensation enabled, this will be part of the typical player experience in the face of network latency. Controlled lab trials should mimic the real-world situation as closely as possible, which means leaving in-game lag-compensation enabled.

7.1.3 Considerations for Creating Artificial Network Conditions

Controlled test environments often look a lot like Figure 7.3, using either Linux + NISTnet [NISTNET, CAR2003] or FreeBSD + dummynet [DUMMYNET, RIZ1997] for the router that actually instantiates desired latency and loss characteristics. Both approaches have been widely tested and utilised, but they must be used with clear understanding of their underlying limitations.

One of the first things to be recognised is that neither FreeBSD nor Linux (in most incarnations) are real-time operating systems. NISTnet and dummynet instantiate bandwidth limits as variable per-packet delays to meet a long-term average rate-cap. They instantiate delays by queuing packets for later transmission. The definition of 'later' is whenever the operating system next schedules the relevant process to check its queues. Thus, an NISTnet-based or dummynet-based 'controlled latency' router may be introducing its own timing errors into the packet streams.

For example, dummynet wakes up and processes queued packets once every 'tick' of the FreeBSD system's software clock. At least up to FreeBSD 5.4 the software clock defaulted to 100 ticks per second. In other words, dummynet's activities are quantised to discrete multiples of 10 ms. Configure a 42 ms delay, and packets will experience between 50–60 ms of delay (relative to their arrival time) depending on how soon the kernel's software clock 'ticks' after each packet's arrival.

Dummynet's accuracy is improved by adjusting the FreeBSD kernel's tick-rate – preferably to at least 1000 ticks per second (one tick every 1 ms, the default for FreeBSD 6.0 onwards). This can be achieved either by recompiling the kernel [FBKERN] or adding the following line to the file '/boot/loader.conf' and rebooting the machine.

```
kern.hz='1000'
```

Modern motherboards will handle this happily and it reduces dummynet's error to 1 ms. Under these circumstances dummynet is quite a useful and accurate tool [VAN2003]. (Linux kernels have a similar issue with default internal tickrates. At the time of writing, some distributors were beginning to ship products with the default tickrate set to 1000 ticks per second.)

Creating artificial packet loss is relatively easy – make a random decision, and either 'drop' the packet or queue it according to whatever latency or rate-cap rules would otherwise be applied.

Artificial jitter, on the other hand, is harder to get right. One might think that it is sufficient to simply randomise the latency added to each packet in a flow of packets. Unfortunately, if per-packet latency is implemented naively, this can result in packet re-ordering. Packet re-ordering is highly undesirable, both in real network paths (where it is possible but rare) and in your controlled network testbed. It can cause TCP to aggressively shrink its flow control window, and cause unpredictable reactions from UDP-based applications.

To understand the potential for re-ordering, consider the following scenario. You have a router capable of adding a fixed latency to packets flowing through it. The fixed latency is being updated (changed) at intervals of tens to hundreds of milliseconds, and set to values randomly distributed between 90 ms and 110 ms. Re-ordering can occur in the following way:

- Packet P1 arrives at time T1, and P1 is assigned a delay of 110 ms.
- The 'jittering' process now modifies the router's fixed latency to 95 ms.
- Packet P2 arrives at time T2, and is assigned a delay of 95 ms.
- Packets P1 and P2 are transmitted (forwarded) when their assigned delay periods expire.
- If T2 – T1 $>$ 15 ms packet P2 will emerge after P1 (normal).
- If T2 – T1 $<$ 15 ms packet P1 will emerge after P2 (re-ordered).

The key is step 3. If the router's artificial delay mechanism transmits packets at a fixed delay after the packet's arrival, independent of packets which have arrived before or after, re-ordering can occur. Versions of NISTnet introduce re-ordering when used to create artificial jitter in the manner described above.

Dummynet may be used to introduce artificial jitter without re-ordering because packets are queued for transmission in the order they arrive [ARM2004B]. [In the example above, P2 would be queued behind, and transmitted after, P1 even if (T2 – T1) was less than 15 ms.]

One final issue with artificial jitter is the choice of randomised fixed-latency values applied to the dummynet-enabled router. If your 'jittering function' uniformly distributes latency, the resulting RTT can end up having a triangular probability distribution rather than uniform. To a statistician this is obvious – the consequence of independent jittering in both directions through the router is the convolution of two uniform distributions, which is a triangular distribution. However, if you use a normal distribution for latency values (probably a more reasonable approximation of real-world jitter) the convolution is also normally distributed [ARM2004B].

7.2 Communication Models, Cheats and Cheat-Mitigation

Wherever a set of rules governs human interactions, some participants will be inclined to ignore, bend or break those rules. Online games are no different in this regard. Human participants bring various personal motivations to the game, not all of which involve playing 'fair' or 'according to the rules'. And when the rules themselves are enforced inadequately or incompletely by the computer, there is a temptation to play by 'what is possible' rather than what is intended. This leads us to the difficult definition of cheating.

Broadly speaking, cheating could be described thus: '*Any behaviour that a player uses to gain an advantage over his peer players or achieve a target in an online game is cheating if, according to the game rules or at the discretion of the game operator (i.e. the game service provider, who is not necessarily the developer of the game), the advantage or the target is one that he is not supposed to have achieved.*' [YAN2005].

Although many cheats do not directly involve the underlying network, all forms of cheating are of interest in the context of online games. When we play human-mediated games face-to-face (for example, a card or dice game), great skill is required to hide cheating from the other participants around the table. When playing LAN-based computer games it is also difficult to hide cheating from your opponents since they will usually be within sight and earshot. However, when playing online, the participants are invariably nowhere near each other and thus unable to verify that everyone's game play experience is consistent with the game's notional rules. Combined with the de-personalising influence of interacting with people through avatars or characters, the temptation to 'cheat' can be substantial.

In this section, we will briefly review how cheating can be classified and then discuss examples of cheats involving misuse of server-side, client-side and network-layer technologies. Finally we will review possible methods of cheat-mitigation.

7.2.1 Classifying and Naming Methods of Cheating

There is no general agreement on classification and naming of cheats, and we will not attempt to force a particular approach here. It has been observed [YAN2005] that cheats and cheating can be described using three orthogonal attributes.

- What is the underlying vulnerability (what is exploited?)
- What are the consequences of cheating (what type of failure can be inflicted?), and
- Who is doing the cheating?

Vulnerabilities can occur in server or client software (bugs), incompletely specified game-state machines (e.g. the game rules programmed into the clients and servers do not cover all possible scenarios), instability of the communication medium (transmission of game traffic can be quite unreliable) or the incomplete hiding of internal game-state information (e.g. ostensibly hidden server-side information leaking to a player via sniffing of network traffic or un-sanctioned modification of the client).

Every cheat brings different benefits and risks to the cheater, influencing the likelihood of a given cheat being used. Risk occurs when the cheat creates a change in game play that is easily noticed by other players, or automated cheat-detection schemes are coupled with effective sanctions against players caught cheating.

Finally, cheats may be implemented by a single player without anyone's knowledge, they may include other players in collusion, or even include trusted nonplayers who have access to the game server or central game-state databases.

7.2.2 Server-side Cheats

Two broad vulnerabilities exist on the server side. First, the server itself may implement game-play rules in a manner that incompletely predicts the space of possible player actions. Second, the human administrator(s) of the game server may not be trustworthy (or be subject to external influence sufficiently to elicit corrupting behaviour).

One example of game-play corruption is the 'escape' cheat. Upon realising that they are losing the game, a player withdraws at the last minute rather than risk having their final score logged for public record and display. Their withdrawal (escape) prevents their own reputation from being damaged and, in some games, denies their opponent a legitimate win.

Escape cheating can occur in any game where the server forgets about a player as soon as they leave. The partial solution involves tracking 'escapees' as well as winners and losers, and publicising a player's ranking in all three categories. Where the game server keeps comprehensive activity logs it may also be possible to reconstruct, and publish, the cheater's lack of success up to the point where they left. This solution works because it modifies the metrics by which 'success' is measured socially, putting pressure on players to be seen to play games through to the end.

For example, at the end of a map, Quake III Arena only prints on screen the scores of players who remained connected to the server. The server logs do contain all kill/death information, so a cheater's actual success can be determined by an external program. Similar problems have been reported with online strategy games such as 'Go' and 'Bridge' [YAN2003, YAN2002].

Alternative methods of pressuring players to stay have their own weaknesses. For example, a person 'escaping' a game may simply be declared the loser if they do not rejoin and continue in a short period. Unfortunately, such a server-side rule invites malicious players to 'frame' another player – use a network DoS attack (discussed later) to force someone else's disconnection in such a way that the victim cannot reconnect in time and thus is declared the loser. (This works better with games such as Go and Bridge, rather than fast-paced FPS games.)

Another example of incompletely specified in-game behaviour was the 'skin cheat' of early Counter-Strike servers. Counter-Strike is a team-play mod of Half-Life, where two competing teams ('terrorists' and 'counter-terrorists') do battle. The original servers did not expect players to request a change in their skin (the image they project to other players) after the beginning of each game. Because it wasn't expected, the server did not block such requests either. So unscrupulous players would switch skins in-game to infiltrate opposition positions, achieving a substantial tactical advantage in the process [CHO2001].

For games that rely on players 'logging in' with a user name and password, there can be issues associated with protection of the password system from attack. The simplest cheat would be to launch a dictionary attack on the online login process of a game server – cycling through thousands of possible words as passwords for your intended victim. On the face of it, the game server can compensate by enforcing a minimum delay to every password attempt and then freeze the account after a certain number of failed login attempts. Unfortunately, this leads to another type of cheat – freeze another player out of the game by deliberately logging into the game server as the victim, using passwords you know are not right. Once you hit the server's limit for bad passwords your victim can also not login. This would be a major problem for the victim if the freeze was triggered just before they were going to login for some critical events in an RTS or MMORPG game.

A broad range of cheating becomes possible when the server administrator(s) cannot be trusted. Administrators can meddle in a 'God' mode within the game, completely subvert the game-state by modifying the game's database from outside the game itself, or modify the game server's internal game-play rules in subtle ways.

One of the authors tried exactly this kind of server-side cheat with a simple Quake III Arena mod. In this FPS, the server tells clients the 'rules' of game physics, and clients fall in line. The author's server-side modification gave extra flight time to rockets fired by the author. Conversely, the rockets of another specific player (the victim) were given truncated flight times (so they often self-destruct after flying a shorter-than-normal distance) proportional to how many rockets the victim had left. (The victim – a relatively good Quake III Arena player at the time – did not notice the imbalance for a number of days.)

The more usual annoyances imposed by server administrators include changing configurable game physics parameters (such as gravity), capriciously banning or kicking off players and generally being a disruptive 'god' character. Such abuses tend to be self-limiting though, as the administrator soon ends up with an unpopular and empty server.

Finally, it is worth noting a particular style of game play, known as *camping*, that is often described as cheating. Particularly in FPS games, there may be places in a map where a player has a good shooting position while at the same time being hard to kill. When so positioned, a player can rack up a large number of kills without moving around (i.e. they 'camp'). On the one hand it may be cheating to use a camping position that exists only because the map was poorly designed. On the other hand, in real life you cannot complain that the enemy is being 'unfair' if they find a location with admirable camping potential – and to the extent a game tries to emulate real-life scenarios, camping is not cheating. The decision ultimately depends on the game players themselves on a case-by-case basis.

7.2.3 Client-side Cheats

A huge degree of trust must be placed in the client side of an online game. The game-play experience is entirely mediated by the combination of game client software and the operating system, and hardware environment on which the game client runs. Unfortunately, this is almost precisely the wrong place to put much trust because the client software runs on physical hardware entirely under control of the player. (Whilst this is trivially true for PC games, it is often also true of console games where the console's internal security mechanisms are reverse-engineered by third parties.)

Cheats in games using peer-to-peer communication models are essentially variations of client-side cheats, made possible because the local rendering of game-state occurs on the player's own equipment.

Most client-side cheats involve manipulating the software context within which the game client operates to augment a player's apparent reflexes and presence, or augment a player's situational awareness by revealing supposedly hidden information.

In FPS games, the most important augmentation is aiming of one's gun. In real-time strategy and role-playing games, augmentation may be in the form of automated 'bots' who play on your behalf to execute basic strategies 24 hours a day while you are off elsewhere (or sleeping). In all interactive game types, there are advantages to being able to execute certain sequences of moves (e.g. jump, spin and shoot). Thus people develop software tools to automate (and thus augment) the player's execution of such moves.

One of the most common augmentations in FPS games is the 'Aimbot' – automated code to assist in aiming and shooting at other players. Once patched into the actual game client software, an aimbot is fully aware of the precise locations of all other players within the client's view. When the player issues 'shoot' commands, the aimbot rewrites the aiming information with precise coordinates of the target, guaranteeing a hit every time. More sophisticated aimbots may track and automatically shoot targets you have pointed at. Naturally, this is considered an extremely annoying cheat by the victims.

Another form of augmentation involves violating the game's laws of physics that would normally impede a player's perception of the world around them. There are a number of examples.

- Over-riding other players' choices of avatar skin (colour schemes), making them clearly stand out visually on screen regardless of ambient lighting
- Revealing tactically useful information on screen about other players (e.g. their remaining strength, what weapons they are carrying), regardless of whether or not the players are actually visible
- Eliminating game elements that obscure a player's vision or knowledge of the surrounding map regions. (E.g. 'fog of war' obscuring maps in RTS games, or snow/rain in most games, and opaque walls and boxes hiding enemies in FPS games.)

These cheats violate the trust model of the original game server and client. The server provides information about the game (e.g. other players or map layout) that would not normally be revealed to the actual player until such time as the player reaches a point in the map where the information would naturally become apparent. Augmentation cheats remove such restrictions.

One source of tactical advantage is to simply augment the skins of other players with a glow or colours that otherwise dramatically contrast with the textures, colours and lighting around the map. This defeats the ability of other players to blend into the shadows and camouflage colouring around a map. To a certain degree you can 'see into' shadows by adjusting the brightness and gamma on your video card and monitor. However, this can create eye strain as the rest of the map becomes too bright and washed out. A client-side software modification targeting only the other players is much more effective. Figure 7.5 shows two versions of the same scene in Quake III Arena with normal and glowing skins. The glow cheat makes the character on the left easy to see even in a fast-paced game.

Another form of cheat involves revealing information about the character in view. Figure 7.6 shows a close-up of a Quake III Arena character with its name and current weapon overlaid (this example also shows the 'glow' hack).

The most useful cheat is the 'wallhack'. This refers to any cheat that allows you to 'see' through opaque objects (such as walls). There is incredible tactical benefit to being aware of an oncoming opponent before they are aware of you. Figure 7.7 shows an example in the context of Wolfenstein Enemy Territory. With the cheat activated, you become aware of the medic and soldier hidden from view at the end of the corridor well before they become aware of you.

Figure 7.5 Scene from Quake III Arena – regular view (top) and 'glow' hack (bottom). Reproduced by permission of Id Software, Inc.

Figure 7.6 Close-up of a Quake III Arena character – regular view (left), name and weapon revealed (right). Reproduced by permission of Id Software, Inc.

Wallhacks can be achieved by modifying game client code directly or modifying external software, such as video device drivers, on which the client relies to render images on screen. Today's graphically advanced games offload much of their actual 3D graphics rendering to the operating system's graphics drivers. Opaque objects are drawn by sending a

Figure 7.7 Scene from Wolfenstein Enemy Territory – regular (top) and with wallhack, glow and names (bottom). Reproduced by permission of Id Software, Inc.

request to the graphics driver to 'draw an opaque object'. The 3D graphics drivers render each scene by drawing more distant objects first and then overlaying these with closer objects. When the closer objects are opaque the more distant objects (e.g. other players) are effectively hidden. If the cheater can modify the graphics driver to interpret 'opaque' as 'mostly transparent' (or 'draw as an open wire frame'), then a wallhack is created without touching the game client software at all.

Client software can sometimes also be tricked into revealing information without modifying or patching the client software itself. Many clients have configurable video display options that players can tweak to optimise their client's relationship with the player's choice of video/graphics card. Sometimes these settings can, themselves, cause discontinuities in how map elements are rendered – creating holes and gaps between walls and doors where they should not be.

Client-side cheats are problematic. Players must trust that every other player's game client is obeying the game's rules and laws of physics, yet the potentially malicious player controls the entire hardware and software environment within which their game client operates. At the end of the day we can only reduce the likelihood of client-side cheats rather than remove them entirely.

7.2.4 Network-layer Cheats

The Internet has a number of characteristics that benefit potential cheaters. There is not a great deal of control over resource consumption – both how much you choose to consume, and how much consumption is imposed upon you. There is not much attention to identity, authentication and privacy at the IP layer either.

Perhaps one of the most annoying cheats is disruption of another player's network connection by 'flooding' it with excess IP packet traffic. Since many players are connected via dial-up or consumer broadband ADSL or Cablemodem links, even a minimalist application of distributed denial of service (DDoS) techniques can be quite effective. The main goal of a DDoS attack is to overload some part of the victim's network access path, leading the victim to experience a large spike in latency and packet loss rates. Depending on how it is applied, the victim may not even consider the degraded service to be unusual. In most cases the victim has no way of blocking the inbound flood of traffic before it reaches, and saturates, the weakest part of the victim's Internet connection.

Figure 7.8 shows how a flooding/DDoS attack could be launched against an unsuspecting victim. In principle, the cheater can launch the attack from anywhere, keeping their own game client's network connection free of excess traffic.

Increasing the victim's latency is intended to lead directly to degraded game play (particularly in FPS games) and eventual disconnection of the victim from the game. The flooding attack may be an end unto itself, or simply a means to trigger a disconnection that will by itself cause negative consequences for the victim (e.g. if the game server

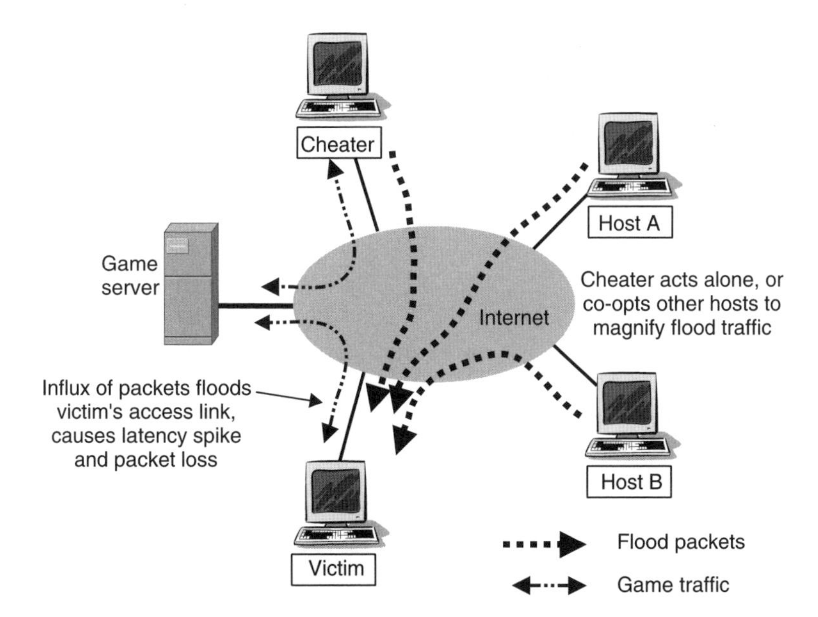

Figure 7.8 DDoS being targeted at a victim while playing a game

penalises disconnections before a game ends). Or the attack could turn the victim into a 'high ping' player, making him or her undesirable to the team on which they had been playing.

This problem is not limited to FPS games. It has been noted that '*... many online Go players choose to play 25 stones in 10 minutes, and it is not unusual for them to play five stones in the last 10 seconds. During the 10 seconds, the* [flood attack] *is always deadly. Most victims lose their games, and they are not sure whether they are attacked or the network connection is jammed due to other normal reasons.*' [YAN2002].

Network traffic is relatively easy to intercept. This truth lends itself to proxies that assist in game-play without requiring any modification to one's game client. Aimbots have been implemented as network-layer proxies on this basis – intercepting and interpreting snapshot packets in one direction and rewriting 'shoot' command packets in the other direction.

Another form of cheating is to connect as multiple users at the same time and 'collude with each other'. For example, one person could play as both partners in a game of online Bridge [YAN2003], thereby having an out-of-band path between the two partners who nominally should have no way to communicate in private.

7.2.5 Cheat-mitigation

A central principle of online, multiplayer game design should be:

- Do not delegate more trust than is absolutely necessary.

There will always be someone in a position to abuse trust. The goal is to spread risk by minimising the amount of control different people have, while still maintaining the desired level of interactivity and immersion/realism. (This is actually a good design principle for all network-mediated communication systems.)

Some server-side cheats are mitigated by discovering (and fixing) incomplete in-game interpretation of game-play rules. Other cheats can be avoided by deploying appropriate personnel security practices with respect to employees who maintain the servers (e.g. authenticated audit trails on database or configuration changes). At the end of the day, server-side cheats are self-correcting – an easily abused server quickly becomes unpopular and either dies through lack of use, or becomes a customer-support nightmare for the hosting company.

Network-layer DoS attacks rely on knowledge of the victim's IP address. Anything that hides player IP addresses will reduce the likelihood of such attacks. This is one significant advantage of client–server network communication model – only the game server needs to know each player's IP address, and each player only needs to know the server's IP address. Games utilising a peer-to-peer communication model at the network layer by definition must expose every player's IP address to every other player.

Note that a game may employ a peer-to-peer system model whilst employing a client–server network model. The peers can utilise a central server at the IP layer to pass game-state updates between themselves without having to reveal their own IP addresses to each other. (Using IP multicast would not help here. Although each player does not need to know anyone else's IP address to send packets, each player's own IP address is revealed in the source address field of packets they send to other clients.)

In principle, a malicious player could attempt to flood through the game itself (e.g. sending in-game text messages to other players at a high rate, which could create the desired flood of traffic at the IP layer). However, a centralised game server can impose limits on the rate at which it forwards in-game messages between players. A reasonable rate limit (on the basis of the ability of the human recipients to read legitimate messages) would be far below the rate needed to flood a victim's underlying network links.

There are two broad approaches to mitigating client-side cheats – minimise the amount of game-state information proactively distributed by the server, and verify that the client's operating environment is within 'normal' parameters.

The following trade-off enables wallhacks to provide tactical advantage:

- There is information a player should not see until 'the right time'.
- To ensure a player sees new events as quickly and smoothly as possible, the server's snapshots carry information about objects and players not yet visible to the player.
- The client software is trusted to only reveal this additional information at an appropriate time and in an appropriate manner.

This set of trade-offs are understandable, particularly as they minimise the processing load at the game server and maximise the client's ability to reveal in-game objects the instant they 'should' come into view.

One solution would be for game servers to only send to each client snapshots pertaining to in-game entities that each player should be able to see. This would negate the value of most wallhacks as the client will not be told what is behind objects that the server believes are opaque. Unfortunately, there are two problems with this solution. First, this increases the processing required to calculate each snapshot at the game server. Second, the server's knowledge of each player's field of view lags by at least one RTT. Thus, the server cannot send data about other objects coming into view until *after* the player has, for example, walked around a corner. This would create a somewhat jarring experience (particularly in FPS games), and likely disadvantage players with high RTT relative to those with low RTT.

A compromise is for the game server to only include in each snapshot the entities the player *can* see and entities the player is likely to see within a very short period into the future (given the trajectories of the relevant entities). The trade-off is this: By reducing how far into the future a server's snapshots 'see' you certainly reduce how much information can be revealed by a wallhack; yet you also reduce the ability of the client to smoothly render other entities coming into view. The converse is also true.

An alternative approach is to verify that the client's operating environment is within 'normal' parameters. This is the principle behind PunkBuster [PBUSTER] and similar services. A third-party application inspects the client software's integrity, client-side configuration variables and the video software/hardware combination to verify that no known 'cheats' are being employed. The cheat-detection application will usually monitor the client host's operating environment continuously throughout a game, rather like an embedded anti-virus tool. For example, PunkBuster (available for 14 FPS games at the time of writing) has both client and server components that communicate regularly throughout a game to ensure that client-side parameters are within bounds as appropriate for the game. PunkBuster's client and server components also regularly 'call home' to the main

PunkBuster site for updates and upgrades, thus automatically staying abreast of the latest cheats.

Current anti-cheat tools are quite sophisticated. Yet, ultimately they too can be corrupted because, as noted earlier, all client-side software runs on physical hardware entirely under control of the player. At best, tools like PunkBuster can only aim to be so complicated and comprehensive that potential cheaters are discouraged from even risking an attempt due to the chance of being caught.

This leads to the question of consequences. Anti-cheat tools are neutered if there are no consequences to being caught. Valve, developers of Half-Life 2 utilise 'Valve Anti-Cheat' ('VAC'), (although the current version is sometimes known as *VAC2*) to detect corruption of client-side environment and explicitly ban players caught cheating. Because of the tight integration of VAC and Valve's Steam player authentication system [STEAM], VAC banning is quite effective. PunkBuster can also keep a record of cheaters, and impose long-term bans if the underlying game's authentication mechanism allows long-term identification of a particular player.

Given the feasibility of introducing severe consequences for cheating, another approach to cheat-mitigation is to verify a game's integrity after cheating has occurred. Although this does not prevent someone from cheating in the first instance, the risk of being later discovered and banned from a game should reduce the incentive for people to try cheating. Proposals along these lines have been made for RTS games [CHA2005]. Players might regularly create cryptographically hashed versions of their map view and own movements and share this information with other players during the game. Clear-text information is only sent when other players claim to have viewable areas overlapping your units. Hashing prevents the other players from extracting information about your units that they should not know at a particular point in time. At the end of the game, each player shares their actual moves and the temporary secret key they used to generate their hashed information. The validity of every other player's claimed viewable area at every step of the game can thus be verified through the (now decoded) log of hashed move information.

More information about cheats and anti-cheat techniques for a range of games can be found in many online forums (for example, Counter-Hack [CHACK]).

References

[ARM2001] G. Armitage, "Sensitivity of Quake3 Players To Network Latency", Poster session, SIGCOMM Internet Measurement Workshop, San Francisco, November 2001.

[ARM2003] G. Armitage, "An Experimental Estimation of Latency Sensitivity in Multiplayer Quake 3", 11th IEEE International Conference on Networks (ICON 2003), Sydney, Australia, September 2003.

[ARM2004A] G. Armitage, L. Stewart, "Limitations of using Real-World, Public Servers to Estimate Jitter Tolerance of First Person Shooter Games", ACM SIGCHI ACE2004 conference, Singapore, June 2004.

[ARM2004B] G. Armitage, P. Branch, L. Stewart, "Mathematical and Experimental Analysis of Limitations in Creating Artificial Jitter for Networked Usability Trials", 3rd Workshop on the Internet, Telecommunications, and Signal Processing (WITSP'04), Adelaide, December 20–22, 2004.

[BEI2004] T. Beigbeder, R. Coughlan, C. Lusher, J. Plunkett, E. Agu, M. Claypool, "The Effects of Loss and Latency on User Performance in Unreal Tournament 2003", ACM SIGCOMM 2004 workshop Netgames'04: Network and System Support for Games, Portland, USA, August 2004.

[CAR2003] M. Carson, D. Santay, "NIST Net: A Linux-based network emulation tool", *ACM Computer Communication Review*, Vol. 33, No. 2, pp. 111–126, 2003.

[CHA2005] C. Chambers, W.C. Feng, W.C. Feng, D. Saha, "Mitigating Information Exposure to Cheaters in Real–Time Strategy Games", Proceedings of the 15th ACM workshop on network and operating systems support for digital audio and video (NOSSDAV 2005), Skamania, Washington, USA, June 2005.

[CHACK] Counter Hack, http://www.counter-hack.net/, Accessed 2006.

[CHO2001] C Choo, "Understanding Cheating in Counterstrike", (online) http://www.fragnetics.com/articles/cscheat/print.html, November 2001.

[DUMMYNET] "Dummynet – Traffic Shaper, Bandwidth Manager and Delay Emulator", http://www.FreeBSD.org/cgi/man.cgi?query=dummynet&sektion=4, Accessed 2006.

[FBKERN] *Chapter 8 Configuring the FreeBSD Kernel*, FreeBSD Handbook, 2005 (http://www.freebsd.org/doc/en_US.ISO8859-1/books/handbook).

[HEN2001] T. Henderson, "Latency and user behaviour on a multiplayer game server", Proceedings of the 3rd International Workshop on Networked Group Communications (NGC), London, UK, November 2001.

[HEN2003] T. Henderson, S. Bhati, "Networked games – a QoS-Sensitive Application for QoS-insensitive Users?" ACM SIGCOMM RIPQoS Workshop 2003, Karlsruhe, Germany, August 2003.

[IPTABLES] "The Netfilter/iptables Project", http://netfilter.samba.org, Accessed 2006.

[MADDEN] "Madden NFL 2004 on EASPORTS.com", http://www.easports.com/games/madden2004/, 2004.

[NIC2004] J. Nichols, M. Claypool, "The Effects of Latency on Online Madden NFL Football", Proceedings of the 14th ACM International Workshop on Network and Operating Systems Support for Digital Audio and Video (NOSSDAV), Kinsale, County Cork, Ireland, June 16–18, 2004.

[NISTNET] National Institute of Standards and Technology, "NIST Net Home Page", USA, http://snad.ncsl.nist.gov/itg/nistnet/, Accessed 2005.

[OLI2003] M. Oliveira and T. Henderson, "What Online Gamers Really Think of the Internet", Proceedings of the 2nd Workshop on Network and System Support for Games (NetGames 2003), Redwood City, CA, USA, May 2003.

[PBUSTER] "PunkBuster Online Countermeasures", http://www.punkbuster.com/, Accessed 2006.

[RIZ1997] L. Rizzo, "Dummynet: A Simple Approach to the Evaluation of Network Protocols", *ACM Computer Communication Review*, Vol. 27, No. 1, pp. 31–41, 1997.

[SHE2003] N. Sheldon, E. Girard, S. Borg, M. Claypool, E. Agu, "The Effect of Latency on User Performance in Warcraft III", Proceedings of the 2nd Workshop on Network and System Support for Games (Netgames 2003), Redwood City, CA, USA, May 2003.

[STEAM] "Welcome to Steam", http://steampowered.com/, Accessed 2006.

[VAN2003] W.A. Vanhonacker, "Evaluation of the FreeBSD Dummynet Network Performance Simulation Tool on a Pentium 4-based Ethernet Bridge", CAIA Technical Report 031202A, December (http://caia.swin.edu.au/reports/031202A/CAIA-TR-031202A.pdf), 2003.

[YAN2002] J. Yan, H-J Choi, "Security Issues in Online Games", *The Electronic Library: International Journal for the Application of Technology in Information Environments*, Vol. 20, No. 2, 2002, Emerald, UK.

[YAN2003] J. Yan, "Security Design in Online Games", Proceedings of the 19th Annual Computer Security Applications Conference (ACSAC'03), *IEEE Computer Society*, Las Vegas, USA, December, 2003.

[YAN2005] J. Yan and B. Randell, "A Systematic Classification of Cheating in Online Games," 4th Workshop on Network & System Support for Games (NetGames'05), Hawthorne, New York, USA, pages 10–11, Oct 2005.

[ZAN2004] S. Zander, G. Armitage, "Empirically Measuring the QoS Sensitivity of Interactive Online Game Players", Australian Telecommunications Networks & Applications Conference 2004, (ATNAC2004), Sydney, Australia, pages 8–10, December 2004.

8

Broadband Access Networks

In this chapter, we examine the networks used to connect users to the Internet. These networks are commonly known as *Access Networks*. There is no single network technology that comprises an access network. Access networks can be based on wireless, coaxial cable, existing telephony cables or fibre optic cable. Access networks might be run by telecommunications companies, corporate entities, private users or a combination of all three. There are many different ways in which users can gain access to the Internet. In this chapter, we survey the most important ones.

8.1 What Broadband Access Networks are and why they Matter

Our interest in this chapter is in broadband access networks. In this context, 'broadband' refers to the bit rate at which the network connects the user. There is no widely accepted agreement as to what bit rate constitutes 'broadband' access. Some definitions have it as a minimum of 256 kbps, others 512 kbps and still others at 1 Mbps or more. For the purpose of this chapter, we will refer to broadband as being at least 256 kbps.

8.1.1 The Role of Broadband Access Networks

Figure 8.1 shows in a simplified form the relationship between the game client, the access network, the Internet and the game server. The client is connected to their Internet Service Provider's (ISP's) router across their access network. Similarly, the game server is connected to their ISP through another access network.

It is important that those involved in developing and deploying games have an understanding of the strengths and weaknesses of the many different kinds of access networks. Different access networks have quite different characteristics, which can have impacts on the quality of the game experience. Depending on the access network, game players may experience quite different effects with regard to latency, jitter, capacity and usage costs.

8.1.2 Characteristics of Broadband Access Networks

Access networks differ from core networks in a number of ways. Generally, these differences are a consequence of the 'Last Mile' problem – the cost of providing connectivity to a geographically distributed population. Core networks shift large volumes of traffic

Networking and Online Games: Understanding and Engineering Multiplayer Internet Games
Grenville Armitage, Mark Claypool, Philip Branch

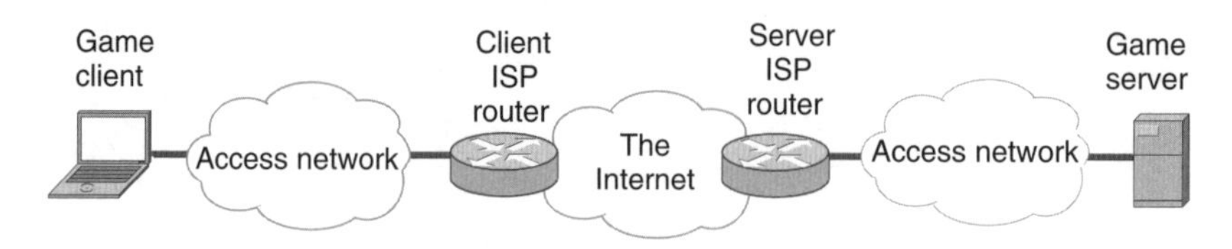

Figure 8.1 Access network

between different nodes in the network, which may be located vast distances apart. Access networks take comparatively smaller traffic volumes from geographically distributed users and deliver it in aggregate form to the core network.

There is no one best solution to connecting users to a core network. The most appropriate solution will depend on many different factors, including the population density of users to be connected, the distance to the nearest core network node (such as a telephone exchange or wireless base station), and existing installed network infrastructure amongst other factors [ARM2000].

Consequently, there is no single broadband access network. Instead, there are a number of different network technologies that come under the general category of being access networks.

Not all kinds of access networks are available in all places at all times. For example, coverage of some broadband wireless networks is often limited. Wireless local area network (LAN) for example, has a range of only tens of metres. Availability of some networks depends on what other infrastructure has been deployed. For example, cable modem networks have usually been deployed in the wake of cable television deployment. Asymmetric Digital Subscriber Link (ADSL) networks are usually deployed on existing two-wire telephony infrastructure. A game developer cannot assume that their player market will either use, or have available for use, a specific access network.

Not all access networks have the same level of reliability. Wireless transmission, although having many advantages over wired transmission, is inherently unreliable [RAPP2002]. Generally, this unreliability is controlled through transmitter and receiver design for constructing a reliable wireless network that provides a satisfactory service to customers, but nevertheless, game players are likely to experience a greater level of unpredictable variations in delay and capacity if their access network is wireless-based than if it is cable based or fibre optic–based.

Usage costs of different broadband networks vary tremendously. The infrastructure needed, the number of users supported by the network and competition from other technologies and service providers, all affect prices that individual users will be charged. Usage costs themselves can be charged in different ways – time-based, volume based, a regular flat fee, or a combination of all three. Although it is difficult to generalise, broadband cellular networks, because of the vast infrastructure investment needed to provide global coverage and seamless mobility support, are typically much more expensive to use than other networks.

Different networks have different capacities. Generally, cellular wireless networks have much lower bit rates than wired networks. Also, some broadband technologies use a shared medium, which means that, although total capacity might be constant, the capacity available to an individual user depends on the number of other users.

Network capacity is not always the same in both directions. The oldest access network application is telephony. With telephony, capacity is usually the same in both directions. Parties involved in a conversation talk, approximately, the same amount during a conversation. Of course, this is not always the case, and some conversations may be quite one-sided. Nevertheless, telephone systems have evolved with the assumption that capacity in both directions needs to be the same. However, many new applications, including most Internet games, require more capacity in the network to the user direction (the forward link) than in the user to the network direction (the reverse link). Typically, a player will receive a great deal of information about the state of the game from the game server on the forward link but will transmit relatively little information (mainly player actions) on the reverse link. Some access networks such as ADSL and cable modem networks naturally support this asymmetry. Their forward link capacity is much greater than their reverse link capacity. Other technologies, such as some cellular broadband networks, are capable of adapting to this asymmetry while other networks provide only symmetric capacity.

Very few access networks are able to offer Quality of Service guarantees. That is, the network guarantees that important connection characteristics such as bit rate, delay and delay variation will be within agreed limits. Specification of these limits form a spectrum ranging from 'best effort' networks where there are no guarantees through networks where guarantees are specified in terms of long-term averages, to networks where bit rates, error rates, delays and variation in delays are tightly specified. As we move across this spectrum, networks generally become less commonly available and more expensive to implement and to use.

Access networks are also often a major source of latency and jitter. Latency is the delay experienced in a game before a command by the user is reflected in the game state. Jitter is the variation in delay. Networks where the same medium is shared among many users, such as wireless and cable modem networks, are more prone to variations in delay than other networks.

Often users will want to access a game while they are mobile. Perhaps they are travelling on public transport, or they might be passengers in a car. Some access networks are able to support mobile users more effectively than others. Although there have been many interesting developments in Mobile Internet Protocol (IP) networks, cellular wireless networks are currently most effective in supporting mobile users.

Some access networks are more convenient to access and to use than others. For example, connecting to a cellular network is usually trivial. The telecommunications company running the network provides direct access to the Internet. However, connecting through a Wireless Local Area Network (WLAN) may require significant reconfiguring of the device used to connect to the network.

In this chapter, we provide an outline of the strengths and weaknesses of the most commonly used access network technologies. We discuss their design, capacity, potential to introduce latency and jitter, reliability, usage costs and how they are typically used. We hope that through understanding these networks, game developers and those deploying games will have a better understanding of the network constraints within which their games must operate.

8.2 Access Network Protocols and Standards

In this section, we discuss some of the general characteristics of access networks.

Access networks are concerned with the delivery of framed packets from the user's network connected device to the network. Access networks are differentiated from each other by their physical and data link layers.

8.2.1 Physical Layer

A network's physical layer specifies the transmission medium and how individual bits are encoded onto it. The media used in access networks are wireless, coaxial cable, twisted copper pairs and optic fibre. The physical layer may also be responsible for minimising or even correcting errors as well as controlling access by multiple users.

Physical media include wireless where communications are mapped onto Radio Frequency electromagnetic radiation. For example, WLANs use a range of frequencies denoted as the Industrial, Scientific and Medical (ISM) Bands of around 2.4 GHz or 5.8 GHz. Cellular networks use frequencies around 800 to 900 MHz or 1900 MHz, depending on the particular network.

Coaxial cable is another commonly used medium. This is most commonly used in cable modems where the access network is overlayed on cable television network infrastructure.

Another broadband access network overlaid on an existing network is ADSL, which uses the twisted pair last hop of the Public Switched Telephony Systems Network (PSTN). Using sophisticated coding and error correction techniques, high bit rates can be obtained over the existing twisted pair medium.

The other commonly used medium is fibre optic cable. With this technology, the medium comprises very thin lengths of glass or plastic down which high-speed lasers or light-emitting diodes send light as the signal. Fibre optic cable is often used in hybrid fibre-coax Cable TV networks.

The physical layer specifies how individual bits are mapped onto the physical medium. The process of mapping a bit or a series of bits onto a medium is referred to as modulation. Wireless systems may encode bits as variations in the amplitude, phase or frequency of a carrier wave. Similar modulation schemes are used for coaxial cable and other electrical wire media. Bits transmitted across a fibre optic cable are typically encoded as pulses of light. The best method of modulation depends on the characteristics of the media it will use and the purpose of the network.

The physical layer specifies how communication channels are mapped onto the physical medium. There needs to be some way of dividing the physical medium into multiple channels. The process of mapping multiple channels onto one physical medium is called *multiplexing*. Multiplexing can be done using frequency division, time division, code division or a combination of all three.

In frequency division multiplexing (FDM), communications are modulated onto a carrier wave of a specified frequency and transmitted. The receiver filters out all frequencies other than that of the specified carrier wave, then demodulates the carrier wave to recover the original signal.

Time division multiplexing (TDM) involves allocating the medium to each channel for short periods of time. These periods of time are commonly called *timeslots*. TDM can be synchronous, where each channel is allocated a fixed number of timeslots in a given period, or asynchronous, where each channel is allocated a variable number of timeslots in a given period.

Code division multiplexing (CDM) uses complex coding to separate multiple channels. CDM is a spread-spectrum technique. In this approach, each user's communication is spread across the available spectrum along with everyone else's but is encoded in such a way as to enable separation of different communication channels. This approach to multiplexing is widely used in wireless communication since it tends to be more resistant to the random variations in signal strength that wireless transmission is commonly subjected to and it makes very efficient use of scarce wireless bandwidth.

8.2.2 Data Link Layer

The data link layer is concerned with transmitting blocks of bits, called *frames*, from the transmitter to the receiver. In cases where multiple users access a single physical channel, it will usually provide some mechanism for managing the resulting contention which may occur if multiple users attempt to access the medium at the same time. This mechanism is identified as belonging to a sublayer within the data link layer, called the *Medium Access Control* (MAC) layer. The data link layer may also be responsible for error detection and correction and perhaps include some mechanism for retransmission of corrupted or missing frames.

Most broadband access networks, with the exception of ADSL use a shared medium and hence have an MAC sublayer. The most commonly used techniques for sharing access are contention-based schemes, where devices wishing to access the medium wait for it to be idle for a period of time before transmitting. If other users also transmit during this time, causing a collision, there must be some mechanism for resolving the conflict and retransmitting. Usually the approach is for all users to wait a random amount of time before retransmitting once the shared medium has become silent. This approach is used in some wired networks such as Ethernet and also some wireless networks such as WLAN. Another approach is a master-slave polling system where one device or node controls access to the medium. It specifies which device may transmit. This approach is also used in some wireless networks, notably Bluetooth and less commonly, WLANs.

The data link layer may also include some form of error correction or error detection, but this too is strongly dependent on the physical medium. In very reliable, high capacity media such as fibre optics, it may be unnecessary to provide any error correction. Where the medium is inherently unreliable, such as wireless, some form of error correction or detection is essential.

The physical and data link layers characterise the Access Network. They specify the medium to be used, how information is encoded, how users are multiplexed onto the medium and if any error control or detection is to be used. We now look at the most significant broadband access networks.

8.3 Cable Networks

Cable television networks are commonly used as an access network where there is a substantial cable television infrastructure. Using cable television networks as an access network allows owners of cable television networks to leverage their investment by providing network access to business and residential customers [ARM2000].

Cable television networks are usually implemented as an inverted tree, with the root of the tree referred to as the *head-end* where programming is injected into the network. The head-end supplies programming to the branches of the tree, which correspond to delivery to particular regions of coverage and localities within those regions, and finally to the leaves of the tree representing the residential receivers of the television signal.

Cable television is built on the existing analogue TV distribution mechanism, and is thus intrinsically an analog medium at the consumer's end. Each television channel occupies approximately 6 MHz. If a coaxial cable network is used from the head-end to the customer premises, then the signal that is distributed occupies Radio Frequencies (RF) in the range of 50 to 450 MHz in the downstream direction. For hybrid fibre-coaxial cable systems that use fibre optic cable in all but the last hop, the frequency range available is much higher, typically from 50 to 750 MHz. In the upstream direction, a much smaller 5 to 42 MHz may be available.

To use this analog medium as an Access Network for transporting digital communication, the digital signal must be modulated onto the analog carrier. An RF signal of 6 MHz can support bit rates in the megabit per second range, but to achieve this, there are a number of obstacles to overcome.

The first difficulty is that cable television is primarily a downstream broadcast medium. Upstream capacity is often limited or nonexistent. The second is that RF transmission is often subject to electrical noise that can severely limit modulation efficiency, particularly in the upstream direction. Finally, even with 750 MHz available, there is not enough capacity to allow a full channel to every user. The capacity has to be shared between multiple users through an MAC sublayer.

Dealing with the situation where there is no reverse channel is the most difficult. In this case, a hybrid solution where downstream communication is provided by the cable network and upstream communication is provided by the telephone network (using a suitable modulation system) is the most common solution, particularly where the network is solely coaxial cable based. Where the network is the more modern hybrid fibre-coaxial network, a reverse channel in the 5 to 42 MHz range, operating at a much lower rate than the downstream network, is a typical solution.

In order to control individual usage, some ISPs impose rate caps on the downstream bit rate. This has been shown to have quite serious consequences for latency when the link is overloaded with data traffic [NGUY2004a]. Even modest rate caps, in conjunction with excess data traffic, can cause latency increases of 100 ms or more. Certainly, this is something that should be of concern to game players and those deploying networked games requiring low latencies.

Standardisation of access networks using cable modems has been dominated by the industry-sponsored Data over Cable Service Interface Specification (DOCSIS). DOCSIS supports the delivery of Ethernet frames between the cable modem and the head-end. It implements a medium access control mechanism allowing shared access by multiple users.

Figure 8.2 shows the main components in a simplified Data over Cable TV network. At the root of the tree that makes up the network is the head-end which provides connectivity to the Internet via a router or bridge. Each user connects to the branch which ultimately connects to the head-end.

Cable networks are an important and effective broadband access network. However, since they are based on a shared medium and rate caps, they can be subject to arbitrary variations in bandwidth and consequent random variations in delay.

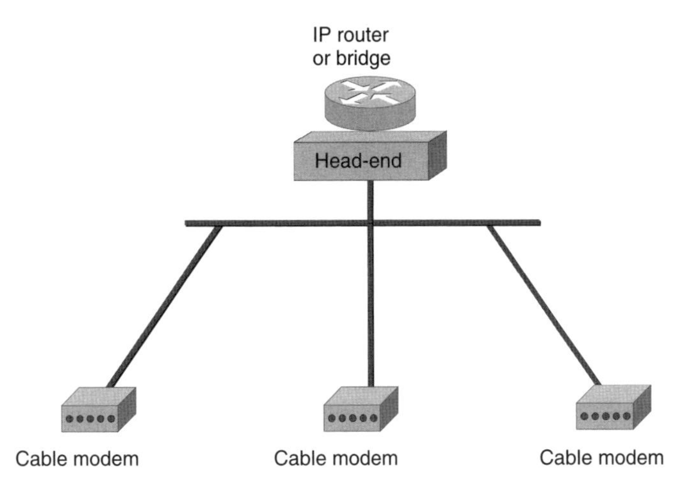

Figure 8.2 Data over a cable TV network

8.4 ADSL Networks

In the same way that cable modems leverage cable television networks to provide broadband access, so Digital Subscriber Line (DSL) technologies leverage existing telephone networks to provide broadband access. The most important of these technologies is Asymmetric Digital Subscriber Link (ADSL). ADSL is a member of a family of subscriber link technologies referred to as xDSL. It includes High-speed Digital Subscriber Line (HDSL), Very high-speed Digital Subscriber Line (VDSL) and others. The most commonly deployed is ADSL [ARM2000].

ADSL uses the existing PSTN twisted pair copper loop otherwise used for standard telephony. It operates in parallel with the existing telephone service but entirely independently of it. Broadband communications are transmitted over the copper loop at different frequencies to that used by standard telephony. With appropriate equipment it can be run in parallel but without affecting the existing telephony service.

Figure 8.3 shows the main components of the ADSL architecture. At each end of the analog loop is an ADSL Transmission Unit (TU). It modulates the digital bitstream onto the local analog loop using frequencies above those used by telephony. At the local PSTN exchange the bitstream is demodulated by another ADSL TU and passed onto the parallel data network via a Digital Subscriber Line Access Module (DSLAM) which provides connectivity to the Internet. Unlike other networks, ADSL is not an end-to-end networking technology. It merely provides the last hop to a customer site. At the exchange, communications over the ADSL link are separated from any other telephony communications and transferred through an entirely independent network.

ADSL is, as its name suggests, highly asymmetric in its upstream and downstream data rates. Downstream rates are typically from 1.5 Mbps to 9 Mbps, whereas upstream rates are typically 16 kbps to 640 kbps. The actual speeds obtained depend on the distance and quality of the copper loop between the customer's residence and the local exchange.

The international telecommunications union (ITU) specification for ADSL (G.992.1) specifies two quite distinct paths with quite different latency characteristics in each direction [ITU1999a]. The fast path minimises latency but at the possible expense of

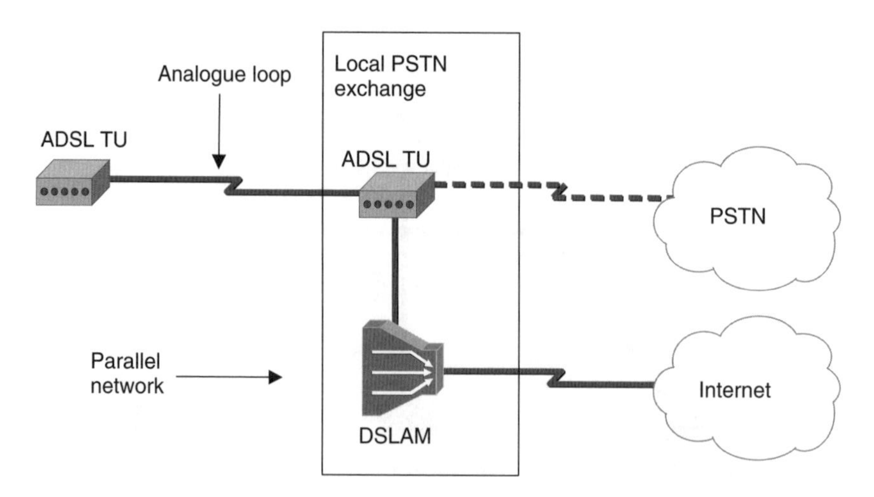

Figure 8.3 Asymmetric digital subscriber line

reliability while the interleave path provides mechanisms for improving reliability but at the cost of additional delay. An interleave path only version of ADSL is defined in G.992.2 [ITU1999b]. In cases where both paths are defined, capacity can be distributed between the two paths. This distribution can occur during use, but causes latency of up to 20 ms.

ADSL is designed to adapt to changing characteristics of the copper loop. Although it occurs infrequently, when link speeds are adapted up or down, quite long delays of up to 3 seconds can be experienced during the adaptation process with obvious consequences for game players.

ADSL is mostly used in association with another access network technology. ADSL provides connectivity between the residence and the exchange. Another home network technology is used to connect individual users to the ADSL network. A typical installation will use ADSL to provide connectivity to the wide area and WLAN to provide multi-user access within the residence. Consequently, apart from variations in latency caused by capacity distribution between the interleave and fast paths, it can also be subject to random variations in capacity caused by sharing the link with other users within the residence.

ADSL is an important broadband technology. It provides high-speed access over existing telephony links. However, like most broadband access networks, it can be subject to random fluctuations in bit rates and latency.

8.5 Wireless LANs

Perhaps the most interesting recent developments in access networks have been in wireless-based access, particularly that class of wireless technologies known as WLAN or WiFi.

WLANs provide great flexibility and convenience for those providing and using an access network. Access network coverage can be provided quickly and easily. Many user devices such as laptops and Personal Digital Assistants now have WLAN networking cards built into them.

WLAN coverage is usually quite small. Depending on the power of the transmitter and receiver, reliable coverage is usually restricted to a radius of twenty metres although with line of sight and sophisticated antennae much greater distances can be covered. However, its purpose is primarily to provide coverage over a reasonably small area. Consequently, WLAN is usually deployed in association with some other access network technology, such as ADSL or cable modem.

8.5.1 IEEE 802.11 Wireless LAN Standards

WLANs use the family of protocols defined by the Institute of Electrical and Electronic Engineers (IEEE) who have also been instrumental in standardising other popular protocols such as Ethernet. WLAN standards are collectively known as the 802.11 standards [IEEE2004], [GAST2002].

802.11 is a family of standards. All 802.11 networks use a common MAC layer but vary in the physical layer details. There have been a number of different physical layer standards released since the original 802.11 standard in 1997, but the most commonly used are the 802.11b and the 802.11g standards operating in the 2.4 GHz ISM band. The ISM band is an area of minimally regulated bandwidth in which anyone may operate radio equipment subject to a minimal set of restraints, primarily on power levels. Consequently, although 802.11 equipment does not need special licensing to install and run, it is potentially subject to interference from other equipment. These include short range Bluetooth communications devices, some cordless telephones and microwave ovens. In designing the IEEE 802.11 protocol, care was taken to make it resistant to interference from other sources. However, this resistance is implemented as a graceful degradation from high bit rates where there is no interference to lower bit rates as interference increases. Consequently, WLANs can be subject to seemingly random changes in bit and error rates when other equipment that broadcasts in the ISM band is used nearby.

802.11b and 802.11g divide the ISM band into 14 overlapping channels with centre frequencies 5 MHz apart. Channels are numbered consecutively from 1 to 14. Within the same coverage area, users may access channels separated by 25 MHz. So in one coverage area users might access channels 1, 6 and 11, while in a neighbouring area they might use 2, 7 and 12. Separation of these channels is important. If user equipment or access points do not maintain this separation, then the result can be additional interference and consequent increases in errors and lower bit rates.

8.5.2 Wireless LAN Architectures

The 802.11 standard specifies two kinds of network architectures: infrastructure and *ad hoc* networks. Infrastructure networks make use of an Access Point (AP) to control communication between users and to provide a communication path to the Internet. In *ad hoc* networks (or peer-to-peer), users' equipment communicates directly without mediation by an AP. For 802.11 to be used as an access network, communication must be via an AP. However, the AP itself needs to be connected to the Internet. This is often via another broadband access network such as an ADSL or a Cable Modem network. Figure 8.4 shows an infrastructure network connecting three APs, each with two wireless nodes. In this kind of network, communication between wireless nodes is usually via the AP rather than directly between them.

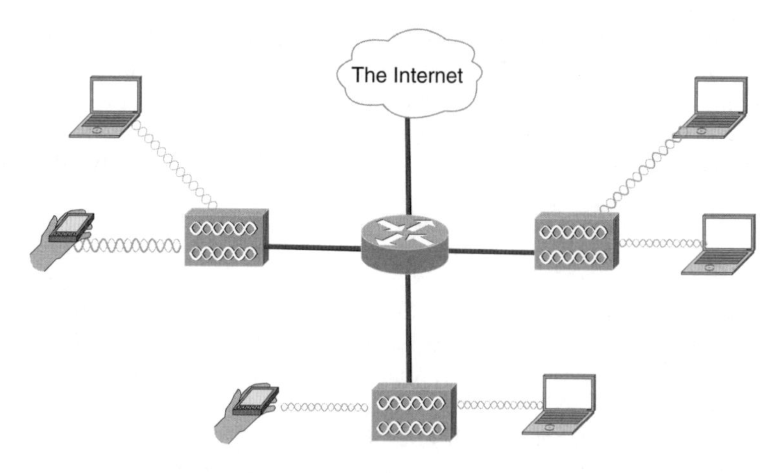

Figure 8.4 Infrastructure wireless LAN

Access to any of the channels supported by the AP is shared. That is, the same channel can support multiple users. However, since the channel has a maximum bit rate, (11 Mbps in 802.11b and 55 Mbps in 802.11g) increasing the number of users decreases the bit rate available to each user.

802.11 networks provide some support for mobility. Users can move from one AP network to another where the APs belong to the same Extended Service Set. However, handover between access points can be quite slow.

WLAN has also experienced security and privacy problems [POTT2002]. Whether this is an important issue for game players is something that individual players and game system operators need to decide for themselves. However, in cases where games intersect with the real economy through professional games with prize money or where game artefacts can be sold for real money, security is certainly important. In general, security on 802.11 networks is weak. The original intention was to have security comparable to wired networks; the so-called *Wired Equivalence Privacy* (WEP). Unfortunately, because it is so easy to eavesdrop on wireless communication, a level of security appropriate for wired communication has proven to be unsatisfactory for wireless communication. Fortunately, the IEEE and the WiFi alliance have recognised this weakness and have released other solutions such as WPA and 802.1X, which provide a more satisfactory level of security.

Shared communication in a wireless environment suffers a problem not experienced in wired communications; that of the hidden terminal. Shared media are usually managed through some kind of contention scheme. A station waits until the shared medium is idle and then, after a random amount of time, transmits onto the shared medium. If another station is also waiting to transmit, there will be a collision. This is usually managed through some random wait before transmitting again. Such a scheme is the basis of Ethernet. However, adapting this scheme to a wireless network is difficult. In a wired medium, a collision is easy to detect. The station monitors the signal it transmits and compares it with what actually appears on the medium. If they differ, then there has been a collision. Unfortunately, detecting collisions where a wireless medium is involved is much more complicated. Two stations may be within range of an AP but not within range of each other. They may both transmit at the same time causing a collision at the AP, but which they will be unaware of. The terminals are 'hidden' from each other.

To manage the hidden terminal problem, WLAN uses Carrier Sense Multiple Access/ Collision Avoidance (CSMA/CA). CSMA/CA works by requiring an acknowledgement from the destination of every frame transmitted. While this is effective in detecting collisions, it has the effect of reducing the available bit rate substantially. Simulations have suggested that for 802.11b, if communication is primarily TCP-based, 5.9 Mbps is available to be shared and if communication is primarily UDP, 7.1 Mbps is available, substantially less than the raw 802.11b bit rate of 11 Mbps.

Most 802.11 networks use the Distributed Coordination Function (DCF), with CSMA/CA to manage contention for the shared channel. In the DCF, a station that wishes to transmit must detect that the medium has been idle for a specified period of time, which is referred to as the Distributed Control Function Interframe Space (DIFS). If another station also tries to transmit at the same time, there will be a collision which will be detected through the CSMA/CA mechanism. The station now waits an Extended Interframe Space (EIFS). The EIFS is a randomly selected length of time whose maximum value increases as the number of failed attempts to transmit increases.

Although DCF is simple, it causes random variations in delay and bit rate. With DCF, once a station gains access to the medium it may keep the medium for as long as it chooses. The number of users sharing the medium and interference from other devices operating in the ISM band will further affect the bit rate available to each user while the CSMA/CA contention mechanisms reduces it even further. Consequently, WLAN access networks can be subject to random variations in delay and bandwidth depending on the number of users and other devices operating in the ISM band.

8.5.3 Recent Developments in WLAN Quality of Service

Because WLAN channels are shared, Quality of Service can be difficult to guarantee. The available capacity is not only shared by all users but also users whose traffic has different round trip times (as the result of one user accessing a local server and another a more remote server) it can experience quite different values in latency [NGUY2004b]. For example, where an 802.11b WLAN is shared between clients accessing data from a local server while others accessing it from a remote server, latencies of up to 100 ms have been observed. For game players, latencies of this order are a serious problem.

To provide some guarantees of Quality of Service (QoS) for delay-sensitive applications, the 802.11e Working group has been investigating enhancements to the 802.11 MAC layer. The work of 802.11e builds upon the little used Point Coordination Function (PCF) defined, but rarely implemented, in the 802.11 MAC layer. The purpose of the PCF is to allow access to the medium in a fair manner controlled by the AP. When PCF is used, time on the medium is divided into a Contention Free Period (CFP) and a Contention Period (CP). During the Contention Period, access is controlled by the DCF. During the Contention Free Period, access is controlled by the PCF. Each station is polled by the AP for any data to be transmitted. If no response is received within a Point Control Function Interframe Spacing (PIFS) then the next station is polled. The PIFS is shorter than the DIFS, ensuring that PCF has priority over DCF. This mechanism provides guaranteed bandwidth for delay-sensitive applications such as voice and games.

Working group 802.11e has developed both the DCF and PCF schemes to introduce a more effective Quality of Service mechanism. The Enhanced Distributed Coordination Function (EDCF) distinguishes between high priority traffic and low priority traffic. High priority traffic waits a shorter period of time than low priority traffic before transmitting. This scheme is simple to implement but provides only relative guarantees of QoS.

The second mechanism developed by 802.11e is the Hybrid Coordination Function (HCF). This is based on the PCF but allows the definition of traffic classes. Different traffic classes can be defined to provide different QoS based on characteristics such as bandwidth and maximum jitter. These are controlled by modifying the frequency with which a station generating a particular traffic class is polled and the maximum number of frames that a station may transmit in response to a poll.

8.6 Cellular Networks

Cellular voice networks such as Global System for Mobile (GSM) and cdmaOne have been commonplace now for over twenty years, but cellular networks able to support packet data at broadband rates are only just becoming widely available. Generally, when compared with other wireless technologies, broadband cellular networks provide much greater coverage, lower latency and seamless handover but at lower bit rates and often, at very high usage costs. Important developments in broadband cellular access networks are Enhanced Data Rates for Global System for Mobile Equipment (EDGE), General Packet Radio System (GPRS), CDMA2000 and Universal Mobile Telecommunications System (UMTS). EDGE and GPRS are sometimes referred to as 2.5G networks, while CDMA2000 and UMTS are sometimes referred to as 3G networks. These technologies can provide broadband wireless access across a much wider area than is possible with other wireless technologies. However, they require the deployment of a great deal of infrastructure, the purchase of bandwidth licenses at often staggeringly high prices, and sophisticated end-user equipment. Consequently, usage costs tend to be high.

Because of their expense and strict bandwidth licensing requirements, cellular networks are usually deployed by large telecommunications companies or specialised wireless communications companies. Comparatively low cost, privately deployed cellular networks are, at least for now, not possible.

8.6.1 GPRS and EDGE

GPRS is derived from the GSM telephony system. Its purpose is to make efficient use of the GSM network for data purposes with a minimum deployment of additional hardware. GPRS is now commonly deployed wherever GSM telephony systems are deployed. However, while GPRS is an excellent packet data network, its delay performance is very poor. It is intended for the efficient communication of nonreal-time communications such as emails and web browsing [PAHL2002]. The GPRS standard defines four delay classes. Class 1 specifies a mean delay of less than two seconds, class 2 a delay of less than 15 seconds, class 3 a delay of less than 75 seconds and class 4 does not specify any delay. Clearly, delays of this magnitude make it unsatisfactory for real-time games. Nevertheless it may be of use to players of nonreal-time games.

EDGE is not so much an alternative network as an improvement to the modulation scheme used in the GSM air interface. By adapting the bit rate to the quality of the link, much higher bit rates are possible than with standard GSM. However, once again its purpose is primarily the transfer of nonreal-time data with no guarantees as to the delay.

8.6.2 3G Networks

3G technologies such as CDMA2000 and the UMTS are probably of much more interest to game players. CDMA2000 has been standardised in the United States by the Telecommunications Industry Authority and UMTS (sometimes W-CDMA) has been standardised by the European Telecommunications Standards Institute (ETSI). CDMA2000 has largely been derived from the earlier IS95 standard (often referred to as cdmaOne) while UMTS is derived from the GSM standard [DAHL1998].

Both UMTS and CDMA2000 support similar services including high-speed Internet access and QoS guarantees. Both can support asymmetric data rates.

Broadband cellular networks are being standardised as part of the International Mobile Telecommunications 2000 Programme (IMT-2000) under the auspices of the ITU. Although terminology differs between standards, broadband cellular networks generally share a common architecture. The simplified architecture for UMTS is shown in Figure 8.5 [UMTS2005]. There are three high-level components in this architecture: the User Equipment (UE), the Universal Terrestrial Radio Access Network (UTRAN) and the Core Network (CN). This high-level architecture is typical of cellular networks [SCHI2003].

In UMTS, the UE is a multimedia capable handset. Apart from voice calls, it supports multimedia calls and packet data. The UE is connected to the UTRAN via the radio interface. The UTRAN is concerned with maintaining connectivity to the UE and with seamless handover both within and between Radio Network Subsystems (RNSs). The CN is responsible for more complex forms of handover and for providing connectivity to other networks, including the Internet.

Each of these systems has an internal structure. The UE is made up of the actual handset itself (the mobile equipment or ME) and a smart card containing user identity

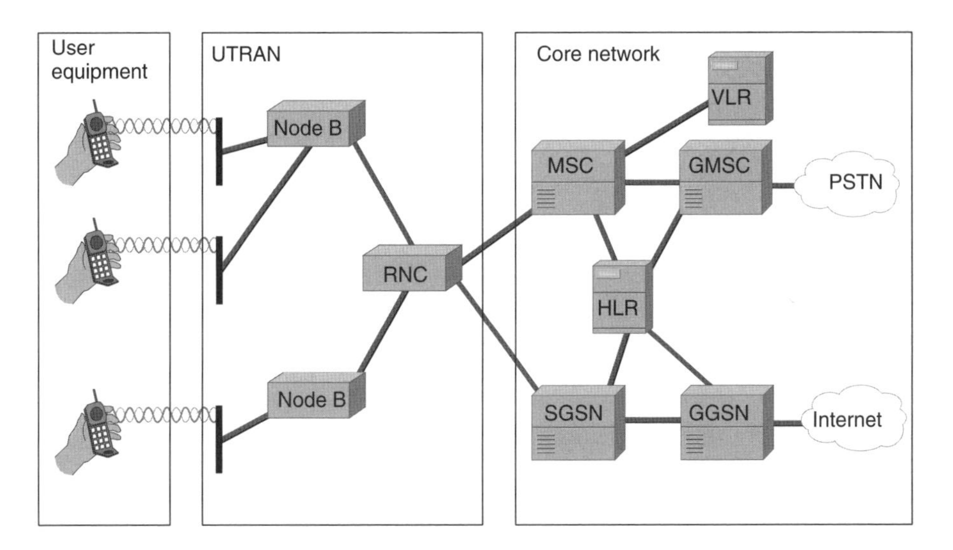

Figure 8.5 UMTS cellular broadband network

information (the UMTS Subscriber Identity Module or USIM). The UTRAN is made up of a number of RNSs. Each RNS contains a number of radio transceivers, perhaps more commonly referred to as *Base Stations*, but known in UMTS as Node B, and a number of Radio Network Controllers (RNCs). Each Node B can have associated multiple radio transceivers. The CN contains even more nodes, the key ones being the Mobile Switching Centre (MSC) responsible for call switching, the Gateway Mobile Switching Centre (GMSC) responsible for connecting voice calls to the Public Switched Telephone Network (PSTN), the Home Location Register (HLR) containing subscriber data, the Visitor Location Register (VLR) containing location data about users, the Serving General Packet Radio System Node (SGSN) which terminates data calls and the Gateway GPRS Support Node (GGSN) which acts as a gateway for packet data to the Internet. From this brief description, it should be apparent that broadband cellular networks are complex and expensive systems.

UMTS defines different levels of service based on two criteria: Quality of Service type and Service Capability. Quality of Service specifies minimum bit rates, delay and delay variation. Although quality of service across the air interface is inherently difficult to provide, the standard allows for flexible allocation of bandwidth to enable a reasonably consistent quality of service to be maintained. The Service Capability specifies a number of capabilities rather than performance requirements. Such capabilities might include location services, perhaps using the Global Positioning System (GPS), extended capabilities built around data stored in the smart card originally included in Mobile Telephones for identification but now, given advances in flash memory capacity, able to store more and different kinds of data, or the Customised Applications for Mobile Network Enhanced Logic (CAMEL) which provides a full development environment.

3G networks are able to support a range of Quality of Service values [3GPP1999]. Latency within the 3G network across the air interface is able to be controlled to a high level of accuracy. Depending on the application, latencies of 10 ms, 20 ms and 80 ms within the network and across the air interface can be specified.

Broadband cellular access networks are reliable, handle mobility well, usually have good coverage and can provide guarantees of quality of service. However, they are also complex, expensive to install and expensive to use. Whether they become a successful and widely used access network is still an open question.

8.7 Bluetooth Networks

Bluetooth networks are included in this chapter even though they are not really an access network technology. Bluetooth is more of a short distance cable replacement technology. Typically, Bluetooth networks extend to no more than 10 metres and are used primarily for peripheral device connectivity such as printers, mice and the like [HAAR1998]. Nevertheless, Bluetooth is an increasingly important technology, typically used in conjunction with broadband access networks or used to construct small *ad hoc* networks. It is perhaps this last characteristic of Bluetooth that is of most interest to game players. Bluetooth-enabled devices can be brought together to form a network with minimal configuration and difficulty.

Bluetooth bit rates are comparatively low. It provides shared bit rates up to 1 Mbps. However, a substantial amount of this is used in Bluetooth overhead leaving approximately 720 to 760 kbps for user data.

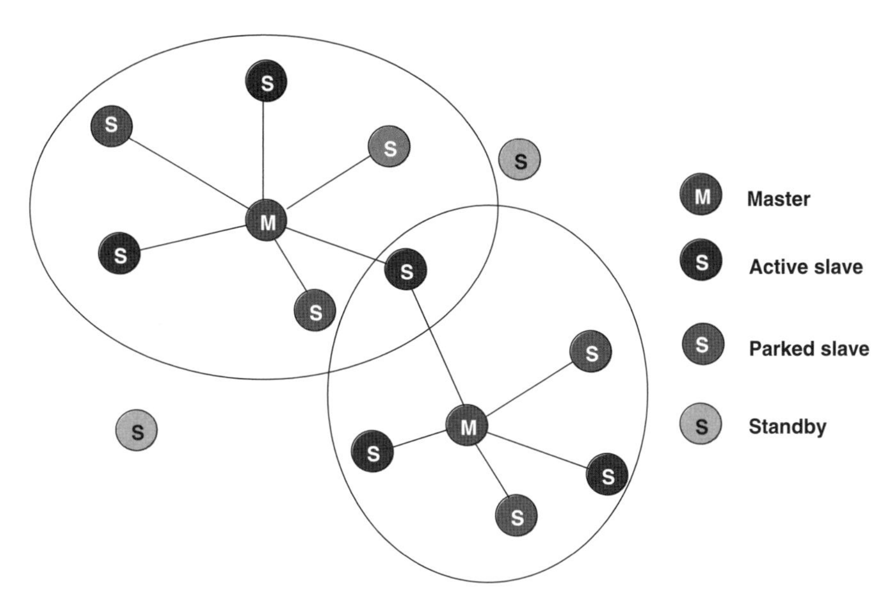

Figure 8.6 Bluetooth scatternet

Bluetooth has been criticised by game players for its latency. Some experiments suggest that typical latency values are between 20 ms and 40 ms; however, higher latencies have been observed [MANS2004]. This is likely to be a consequence of the immaturity of the software rather than a fundamental problem of Bluetooth. In any case, the Human Interface Device profile is claimed to reduce latency to approximately 5 ms.

Capacity is shared through a master–slave polling mechanism. The Bluetooth-enabled devices use an election mechanism to select a master device, which then polls other devices. When a device is polled, it may transmit any frames it has waiting. In this way, bandwidth is shared equitably between individual users. Each master and its slaves forms a piconet with a maximum of eight stations.

Bluetooth uses bridging to link piconets into scatternets. A slave station that forms the bridge between piconets belongs to two piconets. A simplified Bluetooth architecture is shown in Figure 8.6. It shows one master and several slaves in two piconets. One slave which acts as a bridge is a member of both piconets. The two piconets together form a scatternet. Bluetooth supports low power modes of 'parked' and 'standby'.

Bluetooth operates in the unregulated ISM 2.4 GHz band. Consequently, it is subject to interference from other devices that operate in this range. These include 802.11b and 802.11g WLANs, microwave ovens and cordless telephones.

Because Bluetooth is a shared medium subject to random interference, users may experience consequent jitter as other devices start or stop transmitting, or other players join or depart from the network.

8.8 Conclusion

This overview of access networks is intended to give those involved in developing and deploying games an understanding of some of the capabilities and limitations of the many different kinds of access networks.

Broadband access networks offer different levels of capacities, delay, delay variations, coverage, convenience and cost. Those developing and deploying games need to consider the different requirements imposed by these networks. For some access networks, particularly wireless networks, there may be random variations in capacity and delay. Games and game servers must be able to deal with this through buffering or some kind of interpolation. Players using cellular networks may be charged high rates based on traffic volumes. Consequently, game designers need to be careful to minimise the traffic transmitted.

There is a great diversity in the kinds of broadband access networks. This diversity means that most people (at least in developed countries) are able to access some kind of broadband network. But it also means that those developing and deploying Internet-based games need to understand the capabilities and limitations imposed by this diversity.

References

[3GPP1999] 3rd Generation Partnership Project (3GPP);TSG-SA Codec Working Group, "QoS for Speech and Multimedia Codec; Quantitative Performance Evaluation of H.324 Annex C over 3G", 1999.

[ARM2000] G. Armitage, "*Quality of Service in IP Networks: Foundations for a Multi-Service Internet*", Macmillan Technical Publishing, April 2000.

[DAHL1998] E. Dahlman, B. Gudmundson, M. Nilsson, J. Skold, "UMTS/IMT-2000 based on wideband CDMA", *IEEE Communications Magazine*, Vol. 36, No. 9, pp. 70–70, 1998.

[GAST2002] M. Gast, "*802.11 Wireless Networks: The Definitive Guide*", O'Reilly and Associates, 2002.

[HAAR1998] J. Haartsen, "Bluetooth – The Universal Radio Interface for Ad-hoc, Wireless Connectivity", Ericsson Review No. 3, http://www.ericsson.com/, accessed October 2005.

[IEEE2004] IEEE P802.11, "The Working Group for Wireless LANs, the Institute of Electrical and Electronic Engineers", http://www.ieee802.org/11/, accessed October 2005.

[ITU1999a] ITU-T Recommendation G.992.1 "Asymmetrical Digital Subscriber Line (ADSL) Transceivers", June 1999.

[ITU1999b] ITU-T Recommendation G.992.2 "Splitterless Asymmetrical Digital Subscriber Line (ADSL) Transceivers", June 1999.

[MANS2004] K. Mansley, D. Scott, A. Tse, A. Madhavapeddy, "Feedback, Latency, Accuracy: Exploring Tradeoffs in Location-Aware Gaming", SIGCOMM'04 Workshops", Portland, Oregon, August 2004.

[NGUY2004a] T. Nguyen, G. Armitage, "Experimentally Derived Interactions Between TCP Traffic and Service Quality over DOCSIS Cable Links", *Global Internet and Next Generation Networks Symposium, IEEE Globecomm 2004*, Texas, USA, November 2004.

[NGUY2004b] T. Nguyen, G. Armitage, "Quantitative Assessment of IP Service Quality in 802.11b Networks and DOCSIS Networks", Australian Telecommunications Networks & Applications Conference 2004, (ATNAC2004), http://caia.swin.edu.au/pubs/ATNAC04/nguyen-t-armitage-ATNAC2004.pdf, Sydney, Australia, December 8–10, 2004.

[PAHL2002] K. Pahlavan, P. Krishnamurthy, "Principles of Wireless Networks", Prentice-Hall, 2002.

[POTT2002] B. Potter, B. Fleck, "*802.11 Security*", O'Reilly and Associates, 2002.

[RAPP2002] T. Rappaport, "*Wireless Communications: Principles and Practice*", 2nd Edition, Prentice-Hall, 2002.

[SCHI2003] J Schiller, "*Mobile Communications*", 2nd Edition, Addison-Wesley, 2003.

[UMTS2005] UMTS Forum, http://www.umts-forum.org/, accessed October 2005.

9

Where Do Players Come from and When?

In this chapter, we look at various methods for determining where primary player population comes from and when they play. For a game hosting company, it is vital to know the topological and geographical scope of the player population. This information influences location of game servers, provisioning of network capacity, prediction of support costs and targeting of advertising budgets.

Many Internet-based games today use a client–server network communications model (Figure 9.1). By this, we mean that clients scattered all over the world communicate solely with one (or possibly multiple) game servers elsewhere on the Internet. Clients do not exchange Internet Protocol (IP) packets directly with other clients.

An important question to ask is where these clients are typically located and when do they impose traffic on our networks and servers.

In Chapter 7, we discussed how most online interactive games have finite tolerance for latency. A natural consequence is that your player population is most commonly drawn from those locations in the world that are within a tolerable latency of your servers. Think of a game's latency tolerance as the outer radius of a circle encompassing your potential player population. Since (as discussed in Chapter 5) latency can be (very) roughly related to network topology and real-world geography, the tolerable latency radius provides a rough guide to the geographical localities from which your player population may appear.

Knowing where your players are (or potentially could be) coming from helps you target any real-world advertising. It also helps plan new server locations when attempting to cover existing population centres that are not yet served by your game. This knowledge also helps in provisioning link capacity through your Internet Service Provider (ISP), planning ahead for help-desk support calls, and (in conjunction with the issues discussed in Chapter 10) can help in establishing service-level agreements (for IP service quality) with your local and peer ISPs.

In the following sections, we will discuss how to measure game-play and server-discovery usage patterns, and what they reveal about player locations and preferred playing times.

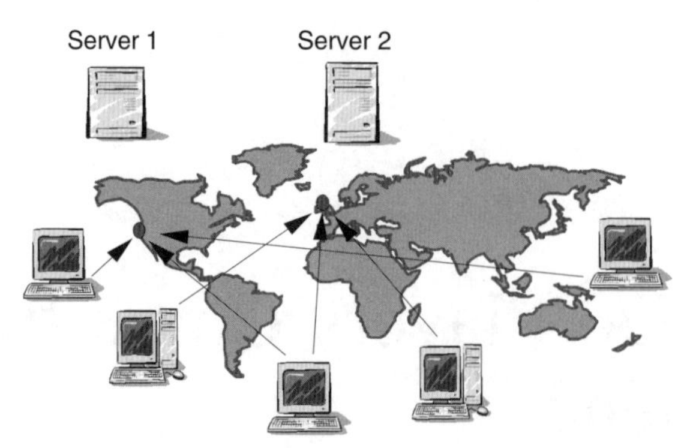

Figure 9.1 Most online games use a client–server network communication model

9.1 Measuring Your Own Game Traffic

There are two main places to measure game traffic – from game server log files, and the network itself. Game server logs provide game-specific data on long-term trends such as client join/leave times and high-level reporting of player activity while on the server. However, server logs do not tell you anything about the actual traffic on your network. For this, you need to deploy packet-level network monitoring equipment. Given the easy availability of free, open-source UNIX-derived operating systems with ethernet monitoring tools, everyone should feel empowered to go and measure real network traffic.

Packet-level network monitoring (often referred to as *packet sniffing*) is simple to do, but not always simple to do precisely. Generally, we want to record the length of every packet, the time every packet is seen (a step known as *timestamping*) and enough of the IP (and UDP or TCP) headers such that we can identify the flow to which every packet belongs. Packet lengths are pretty simple to measure accurately. Complications usually arise when timestamping.

Many of us will want to build a monitoring device from a modern PC with a PCI-bus Ethernet interface and running an open-source UNIX-variant such as FreeBSD [FREEBSD] or Linux [LINUX] (or, conceivably, some version of Microsoft's Windows [WINDOWS]). Unfortunately, generic PC hardware is not generally designed with highly accurate on-board clocks, and these operating systems are not designed for precise, real-time behaviour. As a result, our low-cost general-purpose monitoring system suffers from finite and fluctuating errors in packet arrival timestamping.

Timestamping errors are most often only important for the packet-by-packet traffic patterns discussed later in Chapter 10, so we will postpone discussion of such errors until then. First, let us review how traffic measurement would work in the ideal world.

9.1.1 Sniffing Packets

There are three places we might monitor (*sniff*) packet traffic – at the server, at a client or on a link somewhere in the middle of the network. At the server or client ends, we could track packet flows within the actual server or client host or monitor the directly attached network link with a separate, dedicated host. We would typically want to avoid running packet sniffing software on a game client because the additional processing load may interfere with game-play. Running a packet sniffer on the game server creates similar, but potentially less significant, problems. Ideally, we would use an external device whenever possible.

For external measurements, the first challenge is getting physical access to the packets. Most servers (and many clients) will use switched ethernet as their directly attached link technology. Switched ethernets ensure that each host attached to the network generally only sees ethernet frames going to and from the host itself. In order to see the server's packets, we must interpose an ethernet *hub* (a technological precursor to the ethernet switch) in the path between the server and its network port. Our external sniffing device is then attached to a spare port on the hub, and can see all packets going back and forth.

Unfortunately, ethernet hubs are uncommon for 100 Mbps ('Fast ethernet') links. Placing a hub between the server and its network switch port is likely to force the server (and switch) to run at 10 Mbps instead. An alternative option exists if your network switch (or first hop router) explicitly supports network monitoring by replicating one port's traffic to another, administratively specified port. (Some vendors, such as Cisco, refer to this as 'port mirroring'.) In this case, the server's ethernet link speed is not limited by an interposed 10 Mbps hub. Figure 9.2 shows both of these scenarios.

Having arranged physical access to the ethernet packets, we need an appropriate set of software tools. A good choice would be free, open-source tools such as tcpdump [TCPDUMP] and ethereal [ETHEREAL]. Both allow real-time capture of ethernet frames sent from and received by your host's ethernet interface. Tcpdump is a command line tool whilst ethereal provides a comprehensive GUI. Ethereal provides more comprehensive

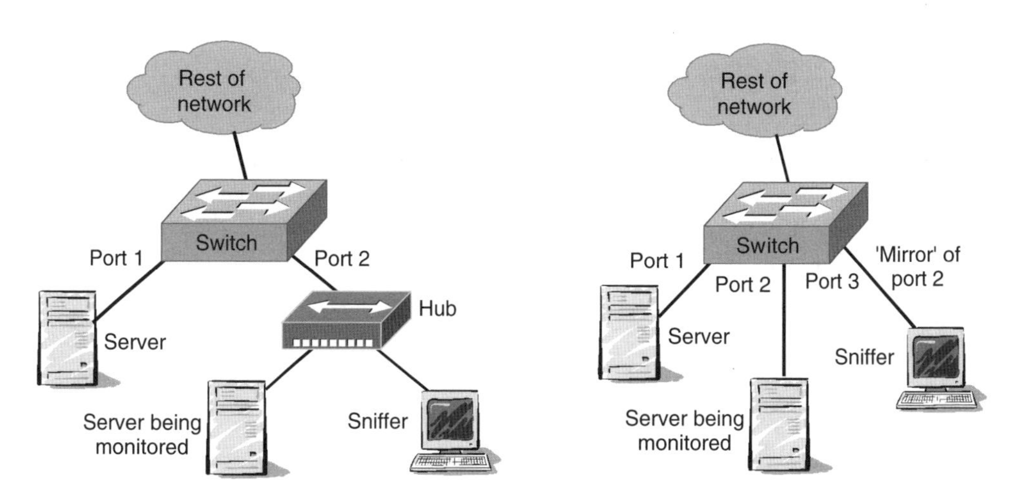

Figure 9.2 Sniffing packets with an external host (using an interposed hub or a switch with port mirroring)

decoding of ethernet frame contents, but for our purposes either tool would suffice. Versions of tcpdump and ethereal exist for most UNIX-derived and Microsoft's Windows operating systems.

(We will not discuss the commercial alternatives – suffice to say that there are many packet sniffer/traffic monitoring applications available.)

9.1.2 Sniffing With Tcpdump

Here is a tcpdump example. Assume we have a standard, possibly old PC (for example, at least an 800 MHz Pentium III) running FreeBSD 5.4 (the production release in late 2005) [FREEBSD]. We have a generic Intel EtherExpress Pro/100+ ethernet card plugged into the PCI bus. The Intel ethernet interface shows up as 'fxp0' (or fxp1, fxp2, etc., if your box has multiple Intel cards).

Figure 9.3 shows the fxp0 interface state as it might be revealed by the console command 'ifconfig'. In this example, fxp0 is the sniffing host's only interface and has IP address 192.168.50.2 on the 192.168.50/24 subnet. The interface appears to be connected to a 10 Mbps hub (the driver as autoselected media type 10baseT/UTP).

Because we are after ethernet frames that are not destined for (or coming from) our local host, the local ethernet interface must run in *promiscuous mode*. Tcpdump will configure this mode by default when run with root user privileges.

The following command:

```
tcpdump -n -i fxp0
```

tells tcpdump to immediately start listening on fxp0, and printing decoded packet trace information to the screen. (The '-n' option speeds things up by suppressing reverse Domain Name System (DNS) lookups of the IP addresses in received IP packets.)

Alternatively, you may want to store a finite number of packets to a file for later processing. The following command:

```
tcpdump -n -i fxp0 -c 200 -w newtracefile
```

tells tcpdump to listen on fxp0 and store the next 200 frames into a file named 'newtracefile'. (Change the '200' to collect more or less frames.) Tcpdump stores the actual ethernet frames in newtracefile, so you can later decode and analyse the traffic at your leisure with:

```
tcpdump -n -r newtracefile
```

(Drop the '-n' option if you want IP addresses converted to their equivalent DNS names.)

```
fxp0: flags=8843<UP,BROADCAST,RUNNING,SIMPLEX,MULTICAST> mtu 1500
        options=8<VLAN_MTU>
        inet 192.168.50.2 netmask 0xffffff00 broadcast 192.168.50.255
        ether 00:03:47:74:73:e7
        media: Ethernet autoselect (10baseT/UTP)
        status: active
```

Figure 9.3 FreeBSD's ifconfig output for fxp0

By default, tcpdump captures and stores only the first 68 bytes of each frame. This is usually enough to analyse the inner IP and UDP or TCP headers. If you wish to capture more of a packet's payload, the '-s NNN' option ensures up to NNN bytes of each packet is captured and saved.

Tcpdump's filter rules come in handy when there is traffic on your network unrelated to the game server under analysis. You can exclude or include particular traffic streams either during or after capture. For example, you might wish to ignore address resolution protocol (ARP) traffic on the local Local Area Network (LAN) during capture:

```
tcpdump -n -i fxp0 -c 200 -w newtracefile not arp
```

or analyse only UDP traffic to or from port 27960 (the default Quake III Arena game port) after capture:

```
tcpdump -n -r newtracefile udp and port 27960
```

Filtering during capture increases the per-frame processing load on the sniffing box, but decreases the disk space required to hold the captured traffic.

Figure 9.4 shows an example of tcpdump's basic text output. In this case, two clients (192.168.23.5 and 192.168.56.90) are playing a game of Quake III Arena (port 27960) on a server at 192.168.50.10. One of the clients appears to be coming in through a Network Address Translation (NAT) box since its source UDP port has been modified to 18756. Timestamps (the left-hand column) provide the hour, minute, second and microsecond at which the packet was seen. On the right is the type and length of each packet.

You can also use tcpdump's '-ttt' option to view inter-packet intervals rather than absolute timestamps.

Filter options can also be used to separate server and client traffic flows during post-capture analysis. For example, the following command:

```
tcpdump -n -r newtracefile src host 192.168.50.10
```

```
                        [...]
12:06:01.434563 IP 192.168.50.10.27960 > 192.168.23.5.18756: UDP, length: 66
12:06:01.462909 IP 192.168.56.90.27960 > 192.168.50.10.27960: UDP, length: 41
12:06:01.476204 IP 192.168.23.5.18756 > 192.168.50.10.27960: UDP, length: 46
12:06:01.482643 IP 192.168.50.10.27960 > 192.168.56.90.27960: UDP, length: 124
12:06:01.482892 IP 192.168.50.10.27960 > 192.168.23.5.18756: UDP, length: 117
12:06:01.488565 IP 192.168.23.5.18756 > 192.168.50.10.27960: UDP, length: 35
12:06:01.504915 IP 192.168.56.90.27960 > 192.168.50.10.27960: UDP, length: 41
12:06:01.534458 IP 192.168.50.10.27960 > 192.168.56.90.27960: UDP, length: 93
12:06:01.534703 IP 192.168.50.10.27960 > 192.168.23.5.18756: UDP, length: 114
12:06:01.546970 IP 192.168.56.90.27960 > 192.168.50.10.27960: UDP, length: 41
12:06:01.547456 IP 192.168.23.5.18756 > 192.168.50.10.27960: UDP, length: 41
12:06:01.580696 IP 192.168.50.10.27960 > 192.168.56.90.27960: UDP, length: 85
12:06:01.581020 IP 192.168.50.10.27960 > 192.168.23.5.18756: UDP, length: 189
                        [...]
```

Figure 9.4 tcpdump output of Quake III Arena traffic in both directions

```
                        [...]
12:06:01.434563 IP 192.168.50.10.27960 > 192.168.23.5.18756: UDP, length: 66
12:06:01.482643 IP 192.168.50.10.27960 > 192.168.56.90.27960: UDP, length: 124
12:06:01.482892 IP 192.168.50.10.27960 > 192.168.23.5.18756: UDP, length: 117
12:06:01.534458 IP 192.168.50.10.27960 > 192.168.56.90.27960: UDP, length: 93
12:06:01.534703 IP 192.168.50.10.27960 > 192.168.23.5.18756: UDP, length: 114
12:06:01.580696 IP 192.168.50.10.27960 > 192.168.56.90.27960: UDP, length: 85
12:06:01.581020 IP 192.168.50.10.27960 > 192.168.23.5.18756: UDP, length: 189
                        [...]
```

Figure 9.5 tcpdump output of Quake III Arena server to client traffic

would (if applied to the traffic in Figure 9.4) only print out the packets from the server to either client (Figure 9.5).

Post-analysis of captured traffic may be done by feeding tcpdump's text output into other software, or writing your own code to parse and interpret tcpdump's on-disk tracefile format. (Tcpdump's tracefiles can be read by many other packet capture programs, including Ethereal.)

9.2 Hourly and Daily Game-play Trends

Long-term game-play trends show us when people tend to play, and thus what times of the day or week we might expect the greatest demands on the latency, jitter and packet loss control mechanisms in our network. If we are hosting a large number of servers, these are also the times when server capacity will be in greatest demand. By tracking and understanding daily and weekly demand cycles, we can estimate user numbers over time, and ultimately estimate bandwidth requirements at the server and client ends of our online games.

9.2.1 An Example Using Quake III Arena

In 2001, one of the authors instrumented two Quake III Arena [QUAKE3] servers in different countries (California, USA and London, UK), and measured the number of players every hour over a period of 3 months [ARM2003]. Figure 9.6 and Figure 9.7 show the daily and weekly cycles respectively for each server. (In Figure 9.7, day 0 is Sunday and Day 6 is Saturday.)

Allowing for the 8-hour time difference between California and London, both servers show usage fluctuations at the same local times. Most players begin joining in the afternoon and on into the evening at each location. This seems intuitively correct, as players would have more free time for online game-play in the late afternoon and evenings.

The time shift of each server's daily cycles reveal player selection of servers with 'better' latency (as discussed in Chapter 7). Both servers were configured identically (same map cycles, same maximum number of players and almost identical server names). When potential players searched through all online servers, only two pieces of information differentiated these two servers – their IP addresses and the round trip time (RTT) to each server. Players self-selected the server with the lowest latency, which tended to be the server topologically closest and in the player's time zone.

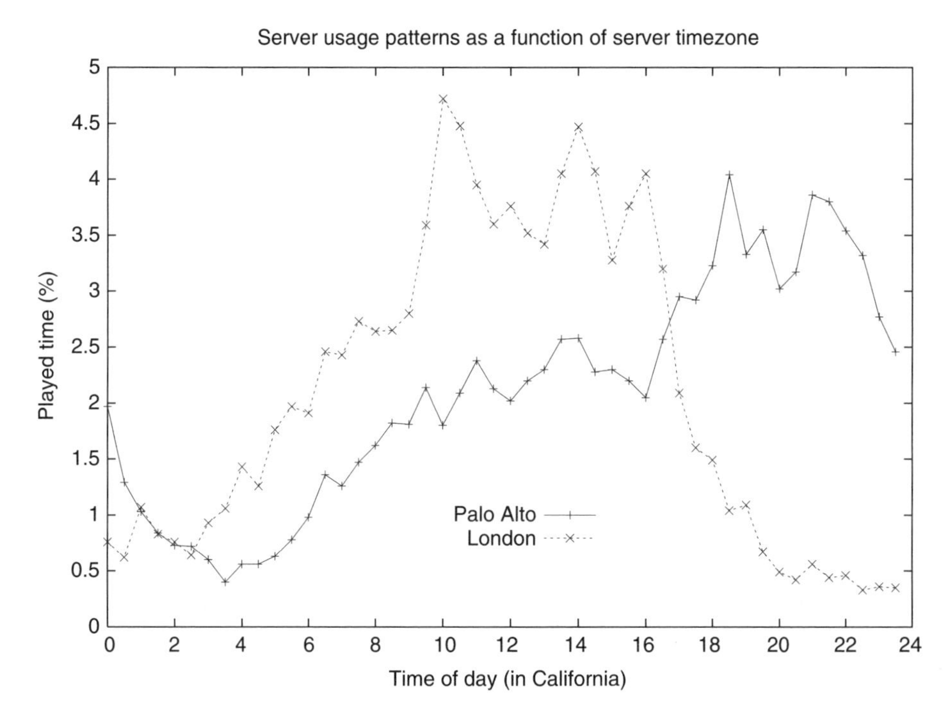

Figure 9.6 Daily Quake III Arena server usage cycles

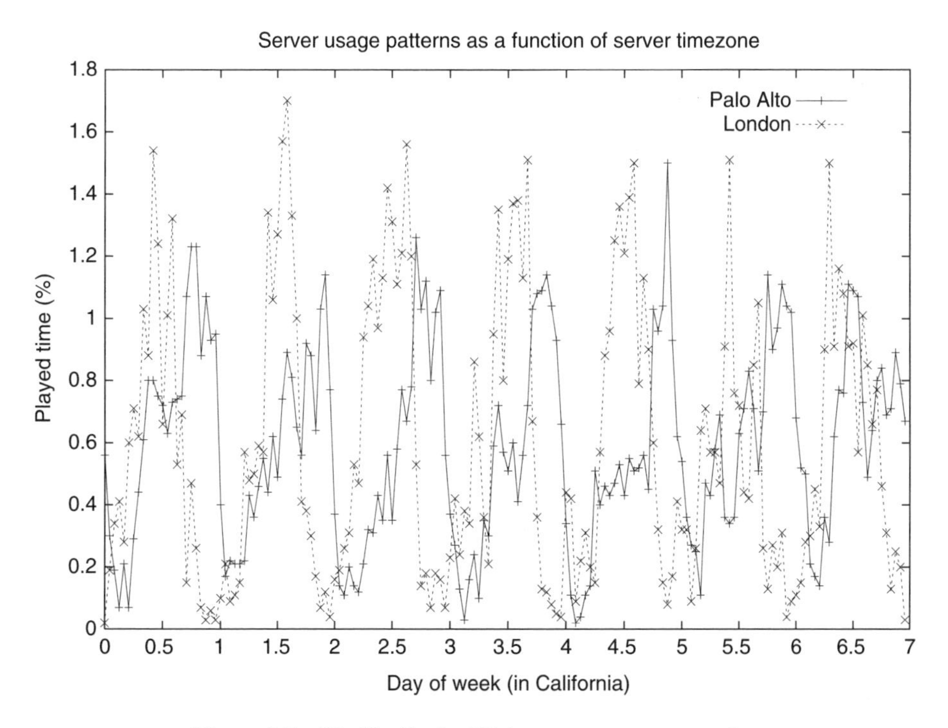

Figure 9.7 Weekly Quake III Arena server usage cycles

9.2.2 An Example Using Wolfenstein Enemy Territory

Figure 9.8 shows broadly similar daily variations experienced by two small Wolfenstein Enemy Territory (ET) servers [WET2005] hosted on the east coast of Australia during late 2004 (data collected over 20 weeks, with each server allowing a maximum of 20 players) [ZANDER2005b]. Usage is revealed by plotting the average number of IP flows associated with actual game play in any given hour of the day. The volume of traffic (in Mbytes) associated with these game-play flows shows the same cyclical fluctuation.

One of the ET servers was located at Swinburne University of Technology (Swinburne) in Melbourne, Australia. The other server was located in Canberra, Australia, on GrangeNet (an experimental, high-speed research network [GRANGE2005]). The CAIA server shows distinct use every afternoon and early evening that drops off by midnight. The GrangeNet server saw only intermittent activity, some days having no players at all. Yet the cycle, such as it is, follows a similar afternoon/evening pattern.

9.2.3 Relationship to Latency Tolerance

As noted in Chapter 7, most players of highly interactive games will come from within a certain 'radius' of tolerable latency. More latency-tolerant games are likely to see their usage cycles spread out somewhat over any given 24-hour period. Latency tolerance translates to a greater number of time zones that might find a given server to be within

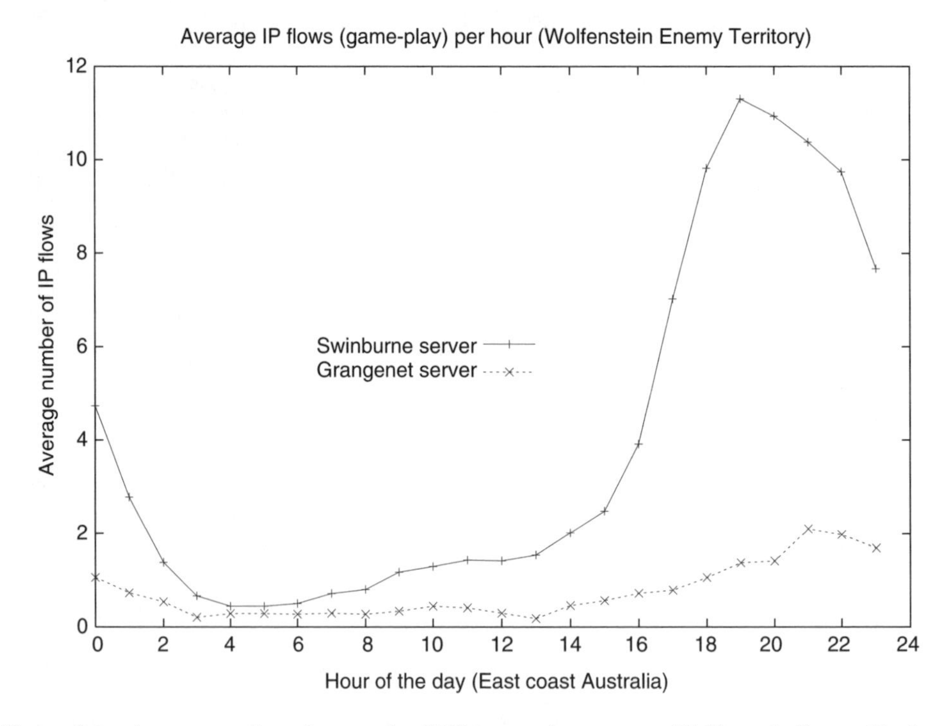

Figure 9.8 Average number of game-play IP flows per hour on two Wolfenstein Enemy Territory servers in Australia

an acceptable latency of the players. Consequently, there is a better chance that someone somewhere might find your server(s) interesting enough to play on regardless of local time at the server.

9.3 Server-discovery (Probe Traffic) Trends

Compared to the traffic generated by actual game play, it is easy to overlook the networkwide traffic caused by in-client and third-party server-discovery mechanisms. This *probe traffic* rises and falls as people turn their game clients on and off, because it derives simply from the act of seeking out servers rather than actually playing on them. It is the 'background microwave radiation' of online game networks, and reveals something about where potential (rather than actual) players reside.

9.3.1 Origins of Probe Traffic

Many First Person Shooter (FPS) games provide an in-game server-discovery function to assist players in finding active game servers on the Internet. The game's publisher typically establishes one or more 'master servers' to hold lists of currently active game servers. The addresses of these master servers are encoded into the game client software.

A player may trigger an automated search process to find game servers of interest. The client queries an appropriate master server, gets back a list of IP addresses representing current game servers, and then proceeds to automatically query each and every server in the list. The queries return information such as server type, current map, number of players and/or teams, number of available player slots, and so on.

Using Wolfenstein ET as the example, Figure 9.9 illustrates how a client discovers active game servers. First, a short UDP packet containing the text 'getservers' is transmitted to *etmaster.idsoftware.com:27950*. This triggers a reply of one or more UDP packets containing 'getserversResponse' followed by a list of ⟨IP address, port number⟩ pairs in

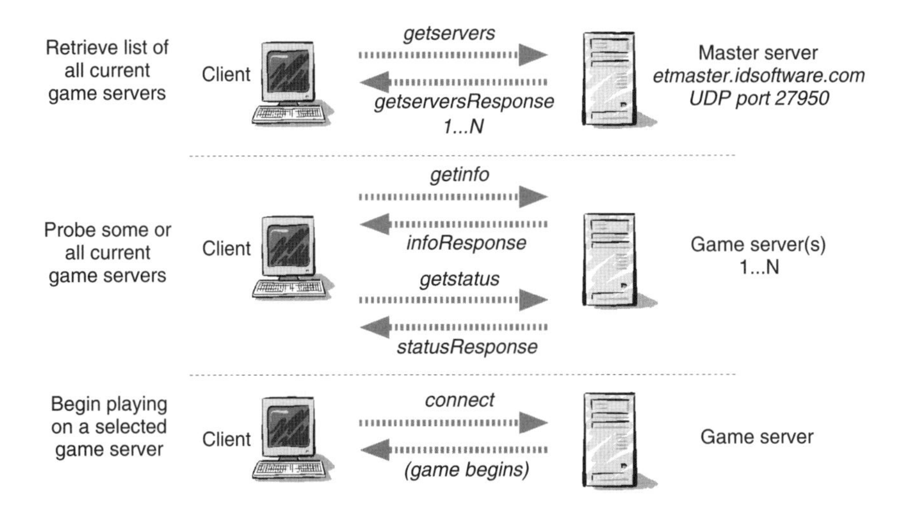

Figure 9.9 Sources of probe traffic: Finding and joining a Wolfenstein Enemy Territory server

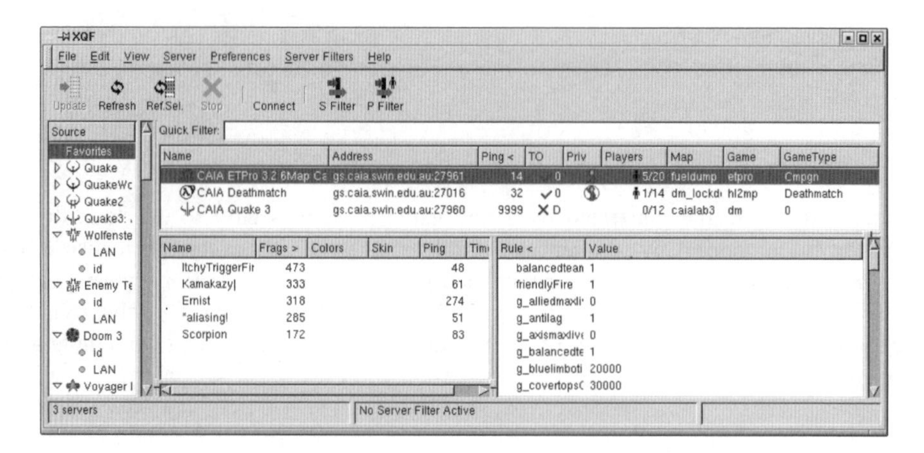

Figure 9.10 XQF – An open-source server-discovery tool

binary format. (These pairs represent the IP addresses and UDP port numbers of remote Wolfenstein ET game servers that have registered with the master server.)

Armed with the getserversResponse list, the client then proceeds to interrogate each listed game server. First, 'getinfo' elicts an 'infoResponse' containing basic server configuration information. Then 'getstatus' elicits a 'statusResponse' containing a detailed list of current players and game information.

The same basic approach is used by other FPS games such as Half-life, Quake III Arena, Half-life 2, etc. Packet sniffing tools (such as tcpdump and ethereal, noted earlier) can be used to observe and dissect the server-discovery packet exchange. (Ethereal has one advantage over tcpdump in this situation – it can decode Quake III Arena/Wolfenstein ET probe traffic.)

An extremely useful tool is the open-source program QStat [QSTAT05]. With Qstat, you can easily probe a wide range of master servers and game servers, archiving the parsed and interpreted replies in text files or piping them to other programs for real-time display. For example, the open-source program 'xqf' [XQF05] provides a GUI front-end for QStat under X11/UNIX and knows about a wide collection of current and old FPS games (Figure 9.10).

9.3.2 Probe Traffic Trends

Although the individual query/response packets are small and the exchange is brief, thousands of clients per day across the planet can create a substantial 'background traffic' over days and weeks.

Some insight into probe traffic patterns can be obtained from the previously mentioned study of two Wolfenstein ET servers in Australia. Figure 9.11 shows the daily cycle of probe traffic seen by the Swinburne server broken out by approximate source region – Europe, Australia, North America, Asia and South America [ZANDER2005b]. (GrangeNet's results were virtually identical.)

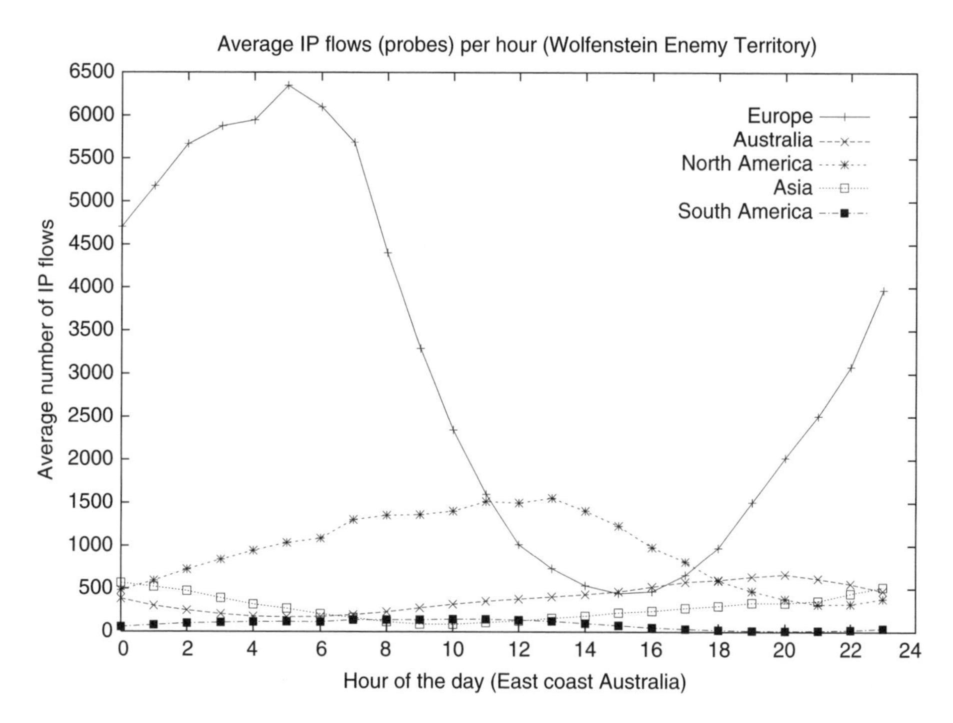

Figure 9.11 Probe traffic per region on Wolfenstein Enemy Territory server in Australia

A number of things are interesting when considered in light of the game-play trends seen in Figure 9.8:

- Both servers saw virtually identical levels of probe traffic despite distinctly different levels of game play.
- Probe traffic trends follow an entirely different time line to the game-play trends.
- The number of probe flows vastly outnumbers the number of game-play flows (by a few orders of magnitude).

This can be explained by observing that probe traffic is dominated by clients who are searching for playable servers, whereas game play is dominated by clients who have found and joined an acceptable server. The former group depends on the demographics of game ownership whilst the latter is more affected by latency, location and server configuration. In Figure 9.11, the probe traffic was dominated by European clients (peaking around 5 am local time), with North American clients coming in distant second (peaking around 1:30 pm local time). Probes from any particular region peak when it is, broadly speaking, afternoon or evening in that particular region.

There is another lesson here. No matter how small you intend your public server to be, you will attract a certain level of probe traffic from around the world simply because you have registered with the master servers. Probe traffic is hard to predict or control locally because it is unrelated to any configuration options on your server itself. (For example, despite their different popularities with actual players the Swinburne and GrangeNet servers each saw around 8 GB of probe traffic across 16 million IP flows during the 20-week period.)

9.4 Mapping Traffic to Player Locations

For marketing and customer-support purposes, it is useful to establish the geographic regions from which your players will (or do) originate. One server-side approach is to map the source IP addresses of clients back to approximate geographic regions (using one of the various 'geo location' services whose databases try to track which ISPs, and hence which locations, have been assigned different IP address ranges). Another approach is to establish the latency tolerance of players on a particular game type (as discussed in Chapter 7), and infer from this an upper limit on the physical distance players are likely to be from your server (a 'latency radius').

9.4.1 Mapping IP Addresses to Geographic Location

Client IP addresses can be extracted from both game play and probe traffic to approximately identify the client's location. In Figure 9.11, the regions were identified using the services of MaxMind's free GeoLite Country Database [MAX2005]. At the time of writing, this resource claimed 97 % accuracy in mapping IP addresses to country codes in their free offering, with higher levels of accuracy and resolution (to the level of states and cities) in their commercial offerings.

A similar service, Geobytes [GEO2005], was used in 2003 to create a world map showing geographic distributions of Half-life: Counter-strike players and servers around the planet [FEN2003]. By calculating the 'great circle' distance between clients (using latitude and longitude information returned by Geobytes) and a specific game server, the authors showed cyclical patterns consistent with the 'people play in their local afternoon and evening' evidence in Figure 9.6 and Figure 9.8.

A further technique is to reverse-lookup client IP addresses into their domain name form. An ISP's own name will often be revealed. In addition, many ISPs embed region-specific codes and names into the domain names of IP addresses attached to customer links. This method was used in [ARM2003] to confirm that players on the Californian and London servers in Figure 9.6 did come largely from ISPs based close to California and London, respectively.

These techniques are likely to be only modestly accurate and not always consistent. An ISP rarely has much commercial incentive for providing enough detailed internal topological knowledge for others to map the ISP's IP addresses to geography. Nor can ISP domain naming schemes be relied upon. Since the names are only required to be useful to the ISP's internal operations staff, there is no guarantee they would use geographically meaningful names or sub-domains. Consequently, IP to country/city mapping databases do, of necessity, include some approximations and guesses.

For example, on the day this chapter was written the author's home broadband connection had been assigned IP address '144.133.92.248'. Doing a reverse DNS lookup returned the domain name (customer premises equipment) 'CPE-144-133-92-248.vic.bigpond.net.au'. We might decode this as CPE device '144-133-92-248' located in Victoria (the 'vic' service region) of the Bigpond ISP (owned by Telstra, Australia's dominant telephone company). The geographic detail is 'vic' and the author was indeed in the city of Melbourne in the state of Victoria at the time. Interestingly, on the same day Geobytes reported this IP address was located in Sydney – about 713 kms away in a different Australian state of New South Wales. (The free MaxMind country database correctly identified the IP address range 144.130/16 to 144.140/16 as being in Australia. But this is far more coarsely grained information.)

9.4.2 Mapping by Latency Tolerance

Establishing latency tolerance requires either controlled lab trials or weeks of monitoring live game-play traffic patterns. (Probe traffic does not help much in this case. We know that someone has probed our server, but we do not know why they did or did not choose to subsequently join.)

Identifying the associated geographical boundaries on your likely player population is, however, non-trivial. A rough approximation would be as follows: Take Z to be some fraction of the speed of light and L to be the latency tolerance of your game. Thus, $R = Z \times L$ is the physical radius from your server that would encompass most of your happy players. We take Z as a fraction of the speed of light because, as discussed in Chapter 5, geographically close players experience additional latency due to convoluted IP layer paths, serialisation delays on slow links, and queuing delays in congested parts of the Internet.

Lab trials have the limitation that you only know a game's latency tolerance, and are left with imprecise methods (such as that described above) to identify geographic areas. Measuring latency tolerance from public server usage patterns is slightly better because you also gain the IP addresses of actual clients who have played. A refined estimation of player location combines latency readings with IP-to-location mapping as discussed above. In addition, armed with the IP addresses of actual clients known to have frequented your server, you can use tools such as traceroute (as discussed in Chapter 5) to measure the paths back to each client. This information, along with time to live (TTL) data revealed in every client's inbound packet to your server, can provide further insights into the relationship between latency tolerance and a player's actual geographical location on the planet.

References

[ARM2003] G. Armitage, "An Experimental Estimation of Latency Sensitivity in Multiplayer Quake 3", 11th IEEE International Conference on Networks (ICON 2033), Sydney, Australia, September 2003.

[ETHEREAL] "Ethereal: A Network Protocol Analyzer", http://www.ethereal.com/ (as of July 2005).

[FEN2003] Wu-chang Feng, Wu-chi Feng, "On the Geographic Distribution of Online Game Servers and Players", In Proceedings of NetGames, 2003, May 2003.

[FREEBSD] The FreeBSD Project, "FreeBSD: The Power to Serve", http://www.freebsd.org (as of July 2005).

[GEO2005] Geobytes, "Geobytes Home Page", http://www.geobytes.com/, 2005.

[GRANGE2005] GrangeNet, http://www.grangenet.net/ (as of July 2005).

[LINUX] Linux Online Inc,"Linux Online!", http://www.linux.org (as of July 2005).

[MAX2005] MaxMind, "GeoLite Country Database IP Country", http://www.maxmind.com/app/geoip_country, 2005.

[QSTAT05] S. Jankowski, "Qstat – Real-time Game Server Status", http://www.qstat.org/, 2005.
[QUAKE3] id Software, "id Software: Quake III Arena", http://www.idsoftware.com/games/quake/quake3-arena/, 2003.
[TCPDUMP] "tcpdump/libpcap", http://www.tcpdump.org (as of July 2005).
[WET2005] Wolfenstein, http://games.activision.com/games/wolfenstein (as of October 2005).
[WINDOWS] Microsoft, http://www.microsoft.com (as of July 2005).
[XQF05] XQF Game Server Browser, http://www.linuxgames.com/xqf/index.shtml, 2005.
[ZANDER2005b] S. Zander, D. Kennedy, G. Armitage, "Dissecting Server-Discovery Traffic Patterns Generated by Multiplayer First Person Shooter Games", *NetGames 2005*, pages 10–11, New York, USA, October 2005.

10

Online Game Traffic Patterns

It is important to understand a number of online game traffic characteristics when planning new network services or trying to improve existing services. As noted in Chapter 9, many Internet-based games today use a client–server network communications model. The Internet Protocol (IP) service experienced by clients of an online game is influenced by the impact of the game's own traffic patterns on the network. Keeping in mind the issues from Chapters 5 and 7, an Internet Service Provider (ISP) who wishes to keep its game playing customers happy requires insights into the competing traffic patterns on its network.

Online games exhibit different patterns and characteristics relative to non–real-time applications (such as email, web surfing and many streaming video/audio applications). Game developers must also consider how their game's communication model translates to actual packet traffic pattern. Where possible, information flow should be smooth rather than bursty, to assist ISPs in managing their infrastructure to deliver better service to game players.

In Chapter 9, we looked at the daily and weekly trends reflecting aggregate join/play/leave cycles of game players themselves. Such statistics help in the provisioning of game servers and long-term sizing of links close to the servers. In this chapter, we will take a closer look at packet-by-packet statistics during game play itself – packet-size distributions and inter-packet arrival times.

These provide an insight into the burstiness of game traffic as perceived by the network, and can be used to model the impact of game traffic on router queues that are being shared with other traffic.

Game traffic can also be divided along the following lines:

- Game-play traffic for which real-time interactivity requirements apply (during a game in progress).
- Signalling/support traffic for which best effort IP service is adequate (for example, server discovery probing, automated map and skin/avatar downloads and patch updates).

In the rest of this chapter we will look at how to measure in-game traffic, provide some examples of traffic patterns from some well-known First Person Shooter (FPS) games and look at how well the game traffic can be simulated and extrapolated from empirical measurements.

10.1 Measuring Game Traffic with Timestamping Errors

As noted in Chapter 9, it is remarkably easy to build free packet-sniffing tools using open-source UNIX-like operating systems and packet-capture software such as ethereal [ETHEREAL] or tcpdump [TCPDUMP]. However, unlike hourly or daily trends, realistic packet-by-packet pattern detection requires sub-millisecond timestamping accuracy. In this section we extend the discussion from Chapter 9 to include the issue of timestamping accuracy with modern, PC-based motherboards.

Most PC-based packet-capture software today will happily report timestamps in a numerical form that implies a resolution of 1 micro-second. The implication is often misleading. As noted earlier, the combination of non–real-time operating systems and PC motherboard clock inaccuracies leads to random (and not so random) variations in reported timestamps from one frame to the next.

There are two key points to remember when calibrating a low-cost PC+software traffic monitor and understanding its limitations:

- Modern PCs have hardware clocks that are not particularly accurate at the micro-second level.
- Operating systems such as FreeBSD [FREEBSD], Linux [LINUX] or Microsoft's Windows [WINDOWS] are not designed to respond to external events (such as packet arrivals) consistently and in predictable time.

Hardware clocks provide the reference against which the software counts the passage of time. They are simply a counter that increments at a fixed, known rate. The operating system relies on knowing how many times the hardware clock increments (*ticks*) per second in order to know how long a second is. If the hardware clock ticks faster than expected, the operating system will overestimate time intervals. Conversely, if the clock ticks slower, the operating system will underestimate time intervals. Since packet-capture programs rely on the operating system for timestamping, the actual timestamps on every packet are subject to vagaries in the sniffing host's on-board clock.

For example, consider a monitoring host whose local reference clock is specified to tick at 1 MHz but, in fact, runs 250 Hz faster (an error of 0.025 %). The operating system will assume that 1,000,000 increments of the clock counter represents one second of elapsed 'real-time', when, in fact, only 0.99,975 seconds have elapsed. Conversely, if packets are arriving over the network precisely 1 second apart, the monitoring software will report that the packets are arriving 1,000,250 micro-seconds apart instead.

Non–real-time operating systems also contribute to inaccuracies by providing no guaranteed response time when handling the arrival of packets. A finite period of time elapses between the ethernet interface hardware receiving a packet and that packet being copied into memory and timestamped. Depending on the operating system's internal architecture for handling I/O (input/output) interrupts, the interval between a packet arriving and being timestamped may depend significantly on the system's processing load at the time. Arriving packets are usually handled 'quickly enough', without any promises from the operating system of how quick that actually is. Consequently, the timestamp attached to every received packet is subject to the vagaries of processor load and interrupt handling at any given moment in time.

Despite all this, we can still use modern PCs and non–real-time operating systems to sniff traffic and report useful inter-packet arrival statistics. There are three things to consider.

- Calibrate your particular combination of hardware and software before putting it to use.
- Minimise unnecessary processor load on the sniffer host.
- Resynchronise the on-board clock regularly.

Calibration involves packet-sniffing a stream of packets from a known, precise source (a number of companies sell precision traffic generators for this purpose) and then comparing the measured spread of inter-packet intervals with the actual inter-packet intervals. This will reveal the bounds of likely error over different timescales and indicate whether a particular hardware and software combination can be trusted.

A trustworthy system might consistently generate timestamps that fluctuate, for example, 15 micro-second around a mean value that is 2 micro-seconds higher than the correct inter-packet interval. We would then attribute ± 15 micro-second error bars, and adjust the means by 2 micro-seconds. Given that many games send packets that are spaced tens of milliseconds apart, this level of accuracy is quite sufficient.

Untrustworthy systems can have all sorts of oddball error modes. For example, we have seen situations where 7 ms deviation in the timestamp was introduced every second or so when an active ethernet cable was plugged into an entirely unrelated interface on the motherboard, or where the operating system only processed (and timestamped) packets from the ethernet card every 10 ms regardless of their arrival time.

Minimising processor load on the sniffer box is another reason it can be undesirable to perform packet capture on the actual game server or client hosts. It is also why you should carefully consider whether to do packet filtering (which incurs slightly more processing load) during or after capturing the traffic.

10.2 Sub-second Characteristics

The impact of network conditions on game traffic is more complex than simply measuring the average bit rate or packet-per-second rate between servers and clients. Many online games have quite low average bit rates. However, after the discussion of serialisation and queuing delays in Chapter 5, it should be clear that packet sizes and inter-packet intervals play an important part in how our game traffic will affect, and be affected by, other IP traffic in a network.

In this section, we will look at traffic examples from a number of FPS games. We focus on the in-game characteristics when players will be most sensitive to network service degradation (such as jitter and packet loss). The packet size and inter-arrival distributions between games (for example, during a map change) will not be covered.

10.2.1 Ticks, Snapshots and Command Updates

During game play, client-to-server transmissions keep the server informed of client actions, while server-to-client transmissions keep the client informed of global state changes in the game as a whole. Precisely how much information is sent, and how frequently, depends both on the game's design and actual game activity at any given point in time.

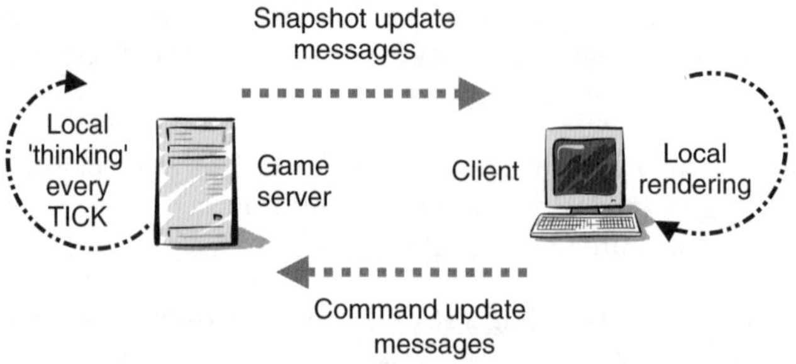

Figure 10.1 Servers and clients 'think' and exchange information at potentially different rates

Figure 10.1 illustrates a useful conceptual model. A server typically 'thinks' in discrete time intervals (which we will refer to as *ticks*). A server must also transmit updates (or *snapshots*) of game state to every client on a regular basis. Similarly, the client is regularly making calculations (rendering game activity on screen) and sending updates (*commands*) to the server.

It is not essential that tick and snapshot rates are equivalent. Most FPS games provide a player-configurable mechanism for the client to request different snapshot rates that are better suited for their downstream network connection (towards the client). Clients can usually also set their command update rate (in packets per second) to suit their upstream network connection.

However, because an FPS server only does things at every tick:

- a client cannot request a snapshot rate faster than the tick rate.
- a client can request a snapshot rate slower than the tick rate, but the result is usually some multiple of the server's tick rate.

An optimal snapshot rate is a trade-off between timeliness of game-state updates and network capacity towards the client. The product of snapshot rate and snapshot packet size is the message rate in bytes per second, which (naturally) must not exceed the available network capacity. If it does, even for brief instances, the client experiences additional jitter (and possibly packet loss).

Designers of online FPS games use various techniques to minimise the size of snapshot packets, such as:

- eliminating redundant or overly precise details in the state variables being sent to clients.
- only sending information about game-state changes within view of the client's current player location.
- only sending the changes between one snapshot and the previous snapshot (sometimes referred to as *delta compression*).

Similarly, commands are kept very small in the client-to-server direction. They need to be sent frequently enough that a player's actions are reflected into the virtual game world accurately. They need to be small because upstream bandwidth is usually highly constrained in dial-up, DSL or cable-modem scenarios.

10.2.2 Controlling Snapshot and Command Rates

Let us take the example of Wolfenstein Enemy Territory (ET) [WET2005]. By default, ET servers tick 20 times per second (a tick every 50 ms) and their clients usually request 20 snapshots per second. If a client requests a higher snapshot rate, such as 30 or 40 per second, the server continues to send snapshots every tick – once every 50 ms. If a client requests 15 snapshots per second, the ET server rounds this up to the next multiple of tick intervals – 100 ms – which corresponds to 10 snapshots per second. The desired snapshot rate can be changed in-game using the client-side console command 'snaps'. (This observation also applies to Quake III Arena [QUAKE3], the engine on which Wolfenstein ET was based.)

Half-life 2 (HL2) [HALFLIFE2004] behaves slightly differently. By default, the HL2 server ticks 66 times per second (once every 15 ms) [HL2NTWK]. As with ET, a HL2 client cannot request more snapshots per second than the HL2 server's tick rate (in this case, 66). However, the situation is rather different when HL2 clients request less than 66 snapshots per second. Rather than simply rounding up to the next multiple of 15 ms, the server mixes transmission intervals to approximate the client's requested snapshot rate over many packets.

For example, a request for 30 snapshots per second results in an uneven mix of packets sent at 30 ms and 45 ms intervals that averages out to 30 packets per second. A request for 50 snapshots per second results in an uneven mix of packets at 15 ms and 30 ms intervals, again averaging out to 50 packets per second. The snapshot rate can be changed in-game by modifying the client-side 'cl_updaterate' variable.

Some games may also allow configurable server-side limits to override the client-side request. For example, setting HL2's 'sv_minupdaterate' server-side variable to 30 would impose a minimum rate of 30 snapshots per second on clients whose 'cl_updaterate' was less than 30.

FPS games usually also allow capping of the downstream server-to-client traffic in bytes per second (not just packets per second). For example, HL2's client-side 'rate' setting allows a client to specify the maximum rate at which it can accept traffic in bytes per second. It might seem redundant to have separate limits on rate (in bytes per second) and snapshots (in packets per second). However, the bytes per second limit protects your downstream link when the snapshot packets become large. If necessary, the snapshot *packet/second* rate will be temporarily reduced if the product of packet length and transmission frequency exceeds the bytes/second rate set by a client.

Command traffic in the client-to-server direction can also be constrained by the player. In Quake III Arena, the client-side variable 'cl_maxpackets' specifies a cap on the number of command packets per second that the client will transmit to the server. This value dictates a lower bound on inter-packet arrival times from client to server.

In the following two sections, we will look in detail at the packet size distributions and inter-packet arrival distributions of some well-known FPS games.

10.3 Sub-second Packet-size Distributions

Not surprisingly, state information sent in server-to-client snapshots is far more varied and complex than the command updates sent from client to server. We will look at some examples from Quake II, Half-life, Quake III Arena, Halo 2, Wolfenstein ET and HL2.

Note that our focus here is on the 'in-game' distributions – packet sizes that are typically seen during the action parts of a game. The examples given here will not include packets seen during idle times and map changes (periods where interactive performance is less important).

Figure 10.2 and Figure 10.3 illustrate typical server-to-client and client-to-server packet-size distributions for Quake II over a Local Area Network (LAN). Figure 10.2 reveals that although individual playing styles can create minute differences in server-to-client packet-size distributions, the overall shape is fairly consistent. Figure 10.3 shows that client-to-server distributions are more clearly bounded regardless of playing style.

Figure 10.4 and Figure 10.5 illustrate similar characteristics for the original Half-life [HALFLIFE][LANG2003]. However, here we see an additional truth about server-to-client traffic – packet sizes depend on map type. In Figure 10.4, four players played on three different maps to show the impact on packet-size distribution. (This should not be surprising – map design influences the frequency with which players interact, and thus influences the amount of information carried by the average snapshot packet.) Over the same three maps (and with four players), Figure 10.5 reveals that client-to-server command packets are not influenced by the map type.

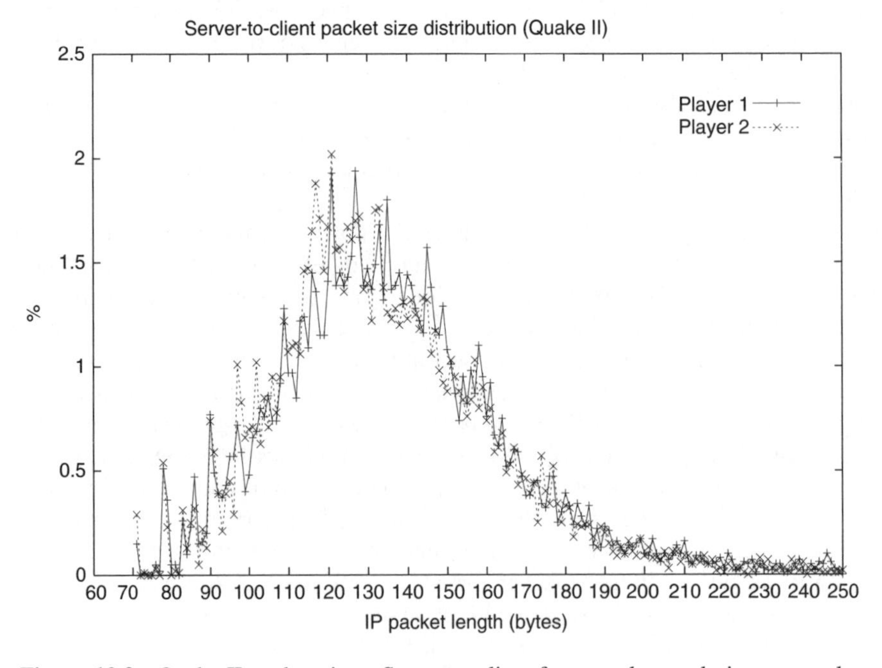

Figure 10.2 Quake II packet sizes: Server-to-client for two players during game play

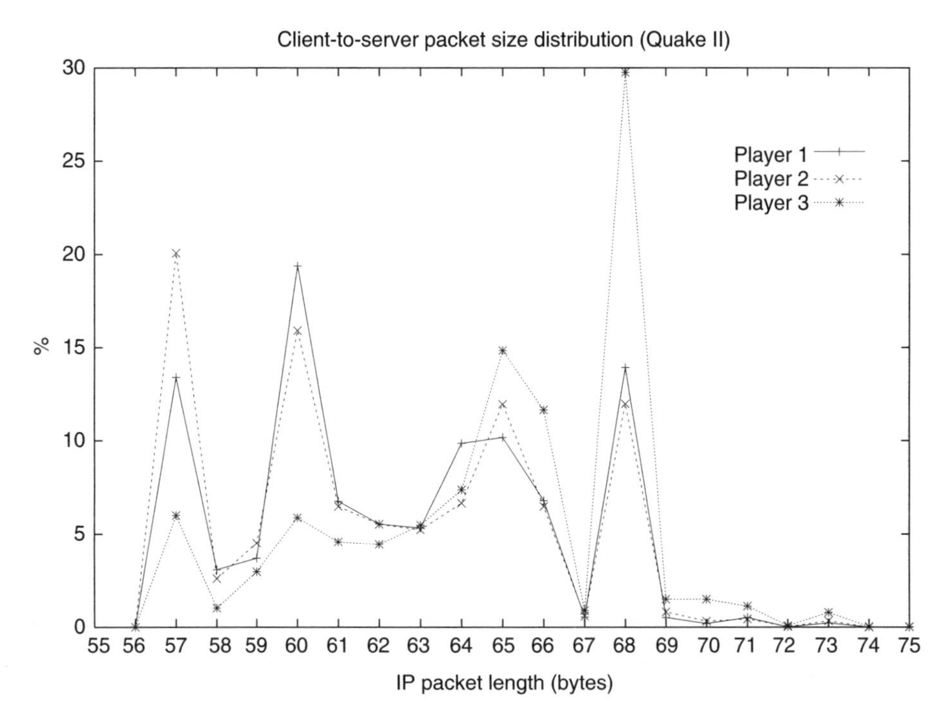

Figure 10.3 Quake II packet sizes: Client-to-server for three players during game play

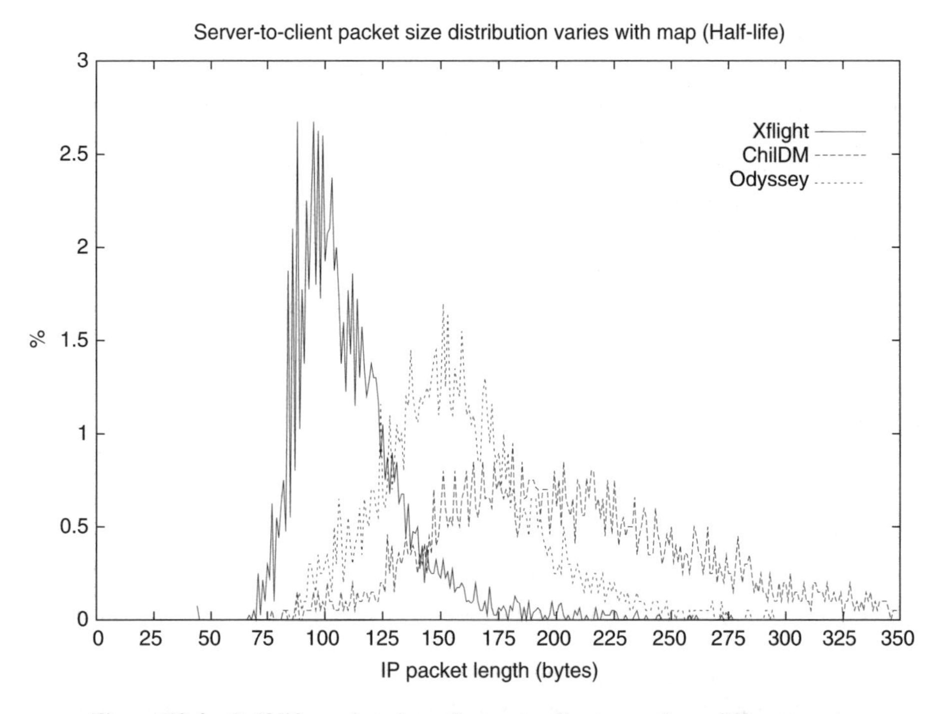

Figure 10.4 Half-life packet sizes: Server-to-client over three different maps

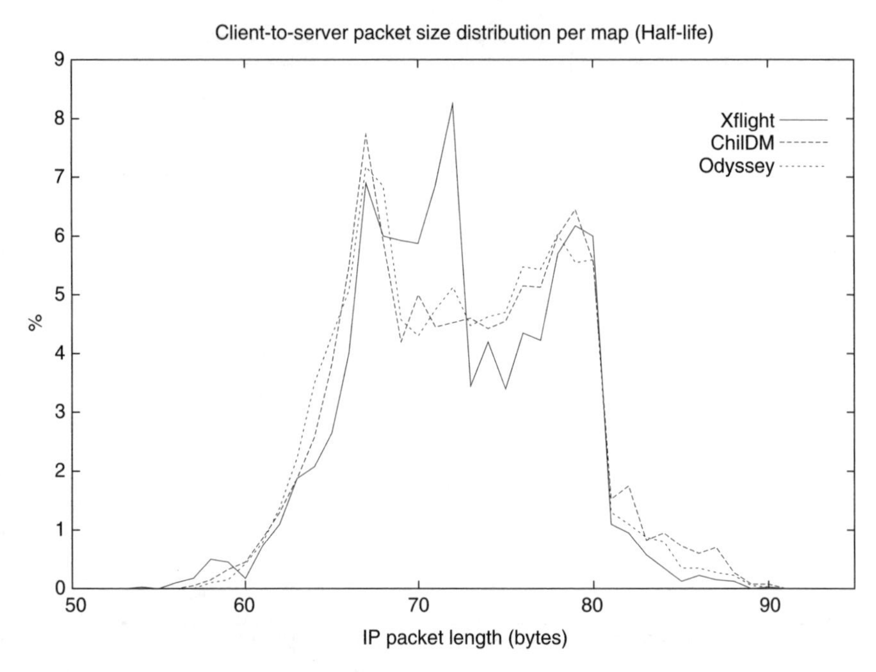

Figure 10.5 Half-life packet sizes: Client-to-server over three different maps

Quake III Arena shows similar client-to-server packet-size distribution curves to Quake II and Half-life in Figure 10.6 (two separate client curves are shown, essentially the same distributions). An additional truth about server-to-client packet-size distributions is illustrated in Figure 10.7 – the snapshots carry more information on average as the number of players goes up. (This example is taken from [LANG2004], where the same map was played multiple times with different numbers of players). Wolfenstein ET client-to-server traffic shows a similar packet-size distribution to that of Quake III Arena.

In general, snapshot size distributions are relatively constant over time for a given map type and number of players. Unfortunately, it is not immediately evident how to predict the likely packet-size distribution of a given map without previous measurements of actual game-play traffic.

A similar set of results emerge from the Xbox game Halo 2 [HALO2004]. Figure 10.8 and Figure 10.9 show the server-to-client and client-to-server packet-size distributions respectively, measured on a LAN when playing in 'System Link' mode [ZANDER2005a]. As expected, the server-to-client distributions increase and broaden out as the number of players increases. However, unlike the earlier PC-based games, a single 'client' (Xbox console) may have one to four players on it. This results in four possible client-to-server distributions – similarly shaped but with progressively larger packets to carry additional player action messages to the server. Halo 2's client-to-server packets only appear in certain discrete sizes, always in multiples of 4 bytes and spaced 8 bytes apart. (The lines in Figure 10.9 visually associate data points rather than imply a continuity of packet sizes between data points.)

Figure 10.10 and Figure 10.11 show the server-to-client and client-to-server packet-size distributions seen with HL2. Since server-to-client size distribution generally varies with

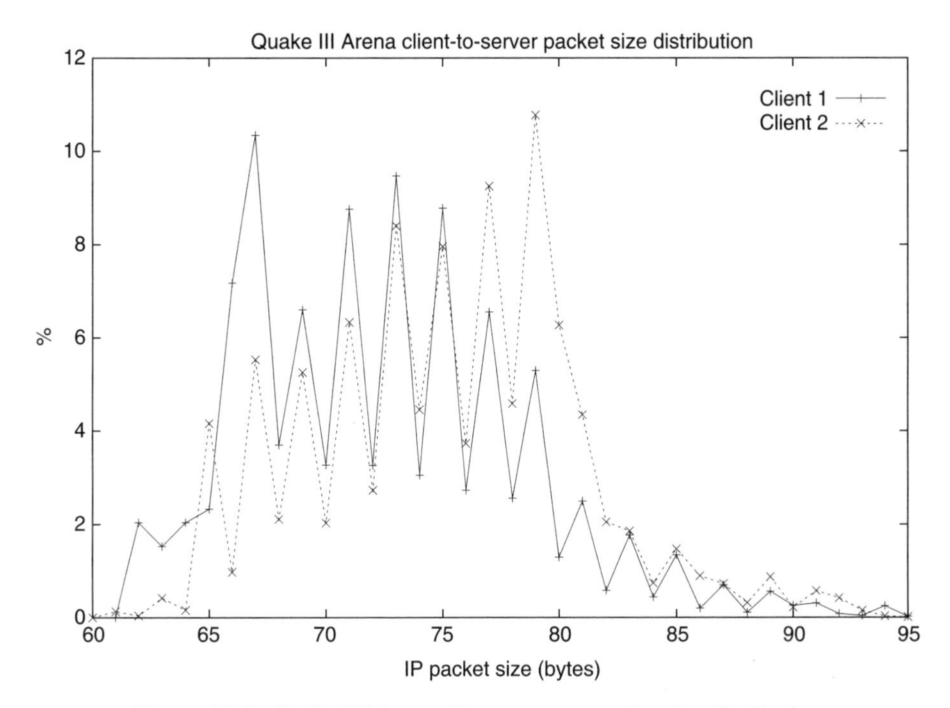

Figure 10.6 Quake III Arena client-to-server packet-size distribution

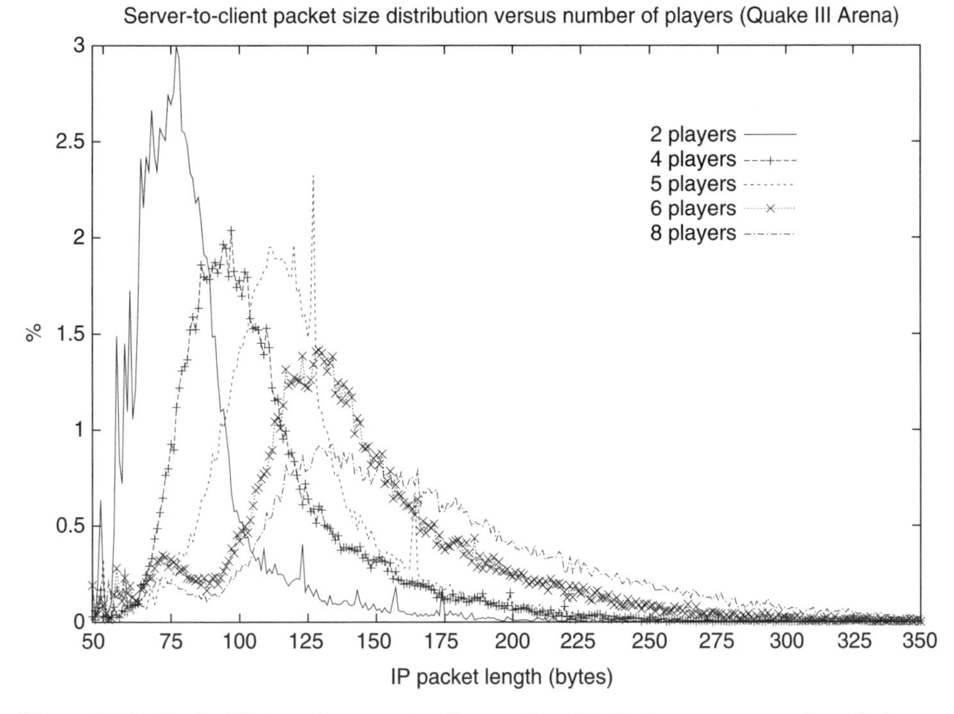

Figure 10.7 Quake III Arena's server-to-client packet distributions versus number of players

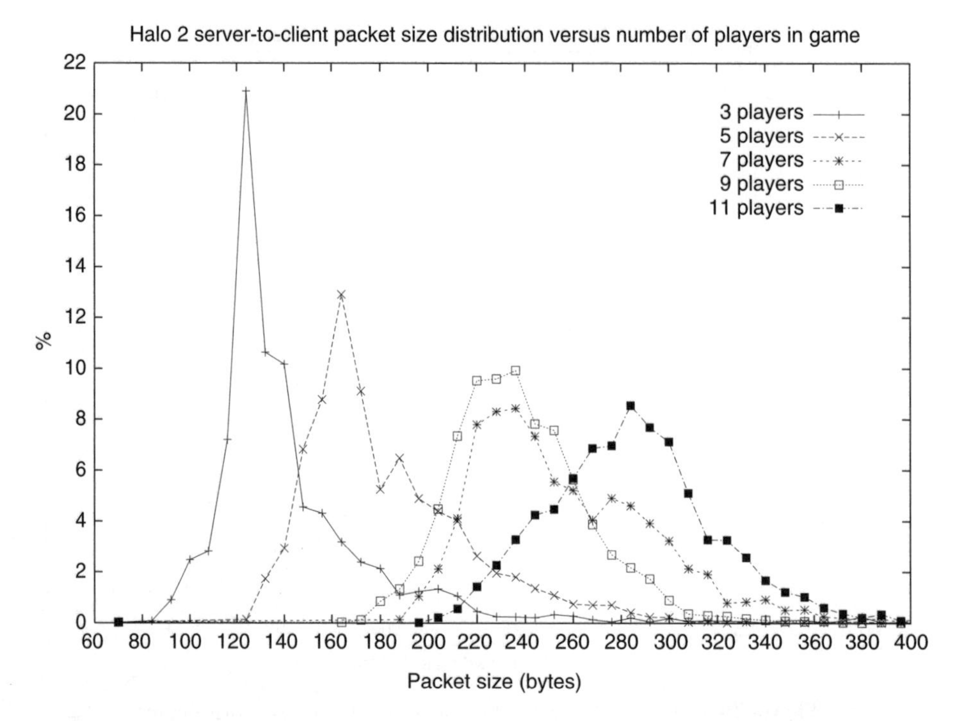

Figure 10.8 Halo 2 server-to-client packet-size distribution (IP packets over LAN)

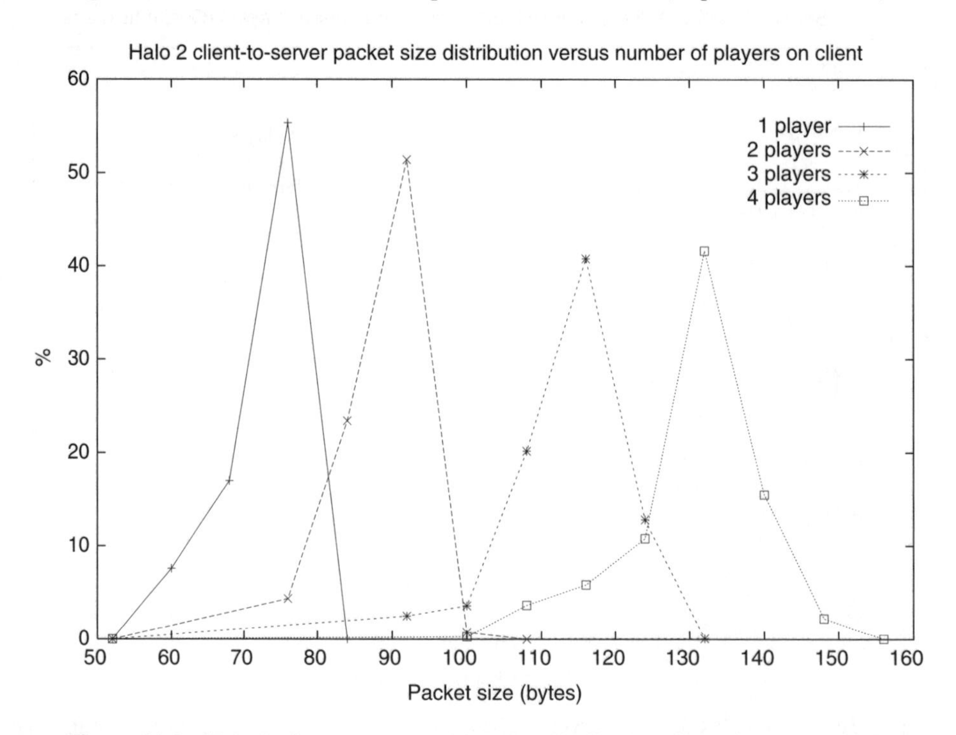

Figure 10.9 Halo 2 client-to-server packet-size distribution (IP packets over LAN)

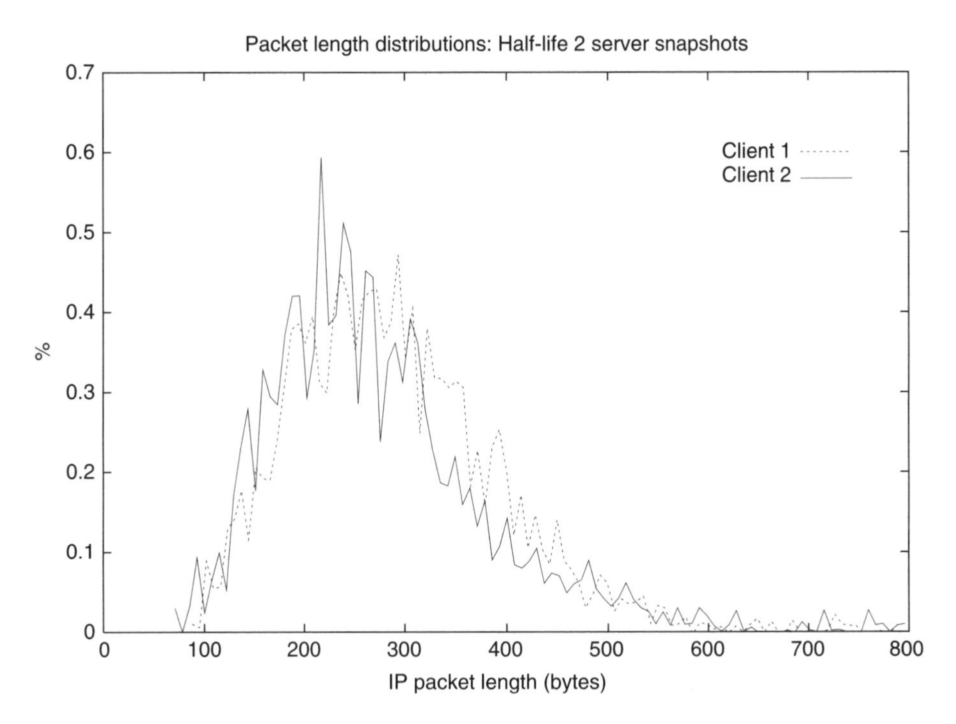

Figure 10.10 Half-life 2 server-to-client packet-size distributions of two clients on the same server

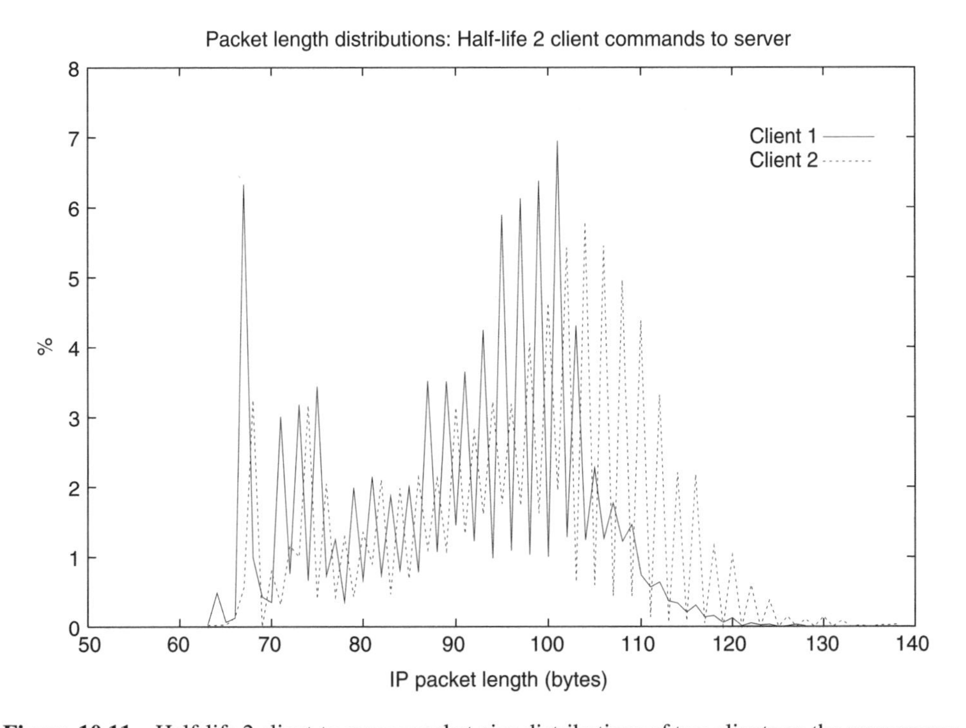

Figure 10.11 Half-life 2 client-to-server packet-size distributions of two clients on the same server

the number of players and map layout, Figure 10.10 should be treated as indicative of what to expect, rather than setting specific boundaries on possible packet sizes.

10.4 Sub-Second Inter-Packet Arrival Times

Inter-packet arrival times reflect the actual, packet-by-packet burstiness of traffic between a game server and its clients. From server to client, the distribution of inter-packet intervals depends on the server's snapshot update algorithm. From client to server, the patterns are more complex, being the result of multiple unsynchronised clients sending player action messages continuously (but somewhat irregularly) to the server.

Once every tick, a game server transmits a back-to-back burst of one or more snapshot packets towards its clients (Figure 10.12). The precise number of packets sent per each tick depends on the snapshot update strategy employed by the server.

Burstiness is highest at the server, gradually dropping as packets branch out along network paths towards each client (Figure 10.13). Inter-packet interval histograms of the aggregate snapshot traffic at a server will show one peak at some multiple of the server's tick interval and another peak down under 1 ms (from snapshots sent back to back during the same tick interval).

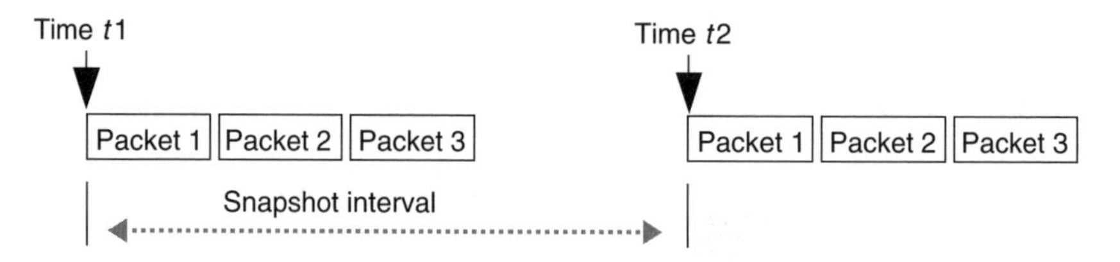

Figure 10.12 Servers transmit one or more back-to-back packets every tick to meet snapshot rate limits

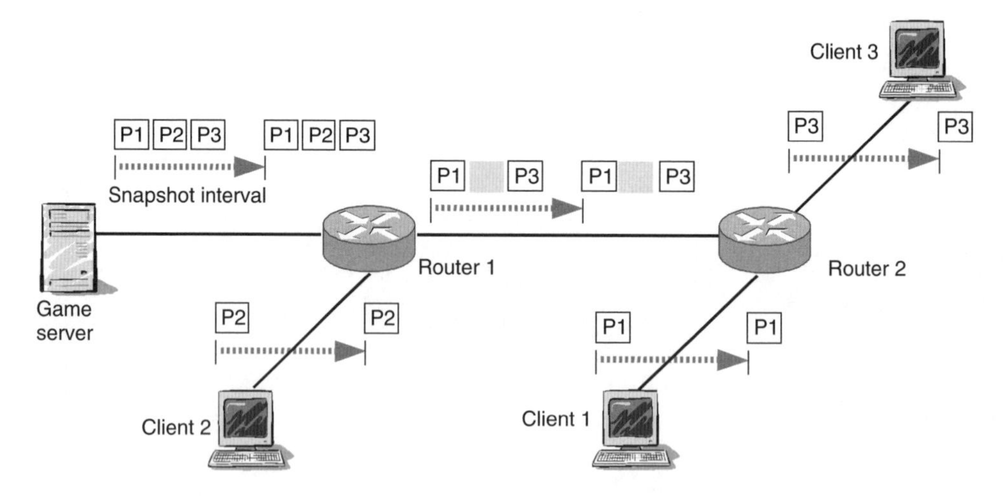

Figure 10.13 Updates become less bursty closer to clients

Table 10.1 Default FPS snapshot and tick rates

Game type	Snapshot rate (per second)	Tick rate (per second)
Quake II	10	?
Half-life	16.67	?
Quake III Arena	20	20
Wolfenstein Enemy Territory	20	20
Half-life 2	20	66
Halo 2	25	?

For a game such as Quake III Arena or Wolfenstein ET, if there are n players on your server, all requesting 20 snapshots per second, then $(100 \times (n - 1)/n)$ percent of the measured inter-packet intervals will be in the sub-millisecond region. The same applies to HL2 if the clients have all requested 33 or 66 snapshots per second. The inter-packet interval distribution becomes more complex to predict when concurrently connected HL2 clients request different snapshot rates.

Table 10.1 shows the default snapshot rates for a number of well-known FPS games along with the server's tick rate (where known).

Snapshot intervals do not generally vary with map choice or number of players. However, game server tick timing can fluctuate by a few milliseconds due to central processing unit (CPU) load, resulting in observable jitter of snapshot transmissions.

Client-to-server intervals of the PC-based FPS games are more difficult to characterise.

- Client transmissions towards the server are uncorrelated (so inter-packet intervals of the aggregate traffic close to the server are spread widely).
- Many games have user-configurable rates (and thus minimum inter-packet interval) at which they send updates from client to server.
- Actual distributions seem to depend strongly on each player's game-play behaviour, the client settings and their system's technical capabilities (e.g. CPU speed and graphics card hardware acceleration) [LANG2003, LANG2004].

Most published empirical results suggest intervals clustered between 10 ms and 30–50 ms. (Although this is not always so – Halo 2's XBox System Link clients transmit their updates at regular 40 ms intervals, same as the server-to-client inter-packet interval.) [ZANDER2005a]

One final note. When interpreting inter-packet interval statistics, be aware that you may sometimes find multiple clients hiding behind a single IP address. Two or more people playing on your Internet-based server from their home LAN will usually be connecting through a home 'gateway' implementing Network Address Translation (NAT). Their user datagram protocol (UDP) packets will have the same home IP address. It is important to always use both the port number and IP address to differentiate remote clients. (You may also be able to confirm multiple distinct client machines behind the NAT box by the unique 'IP ID' field patterns in packets from each client [BELLOVIN2002].)

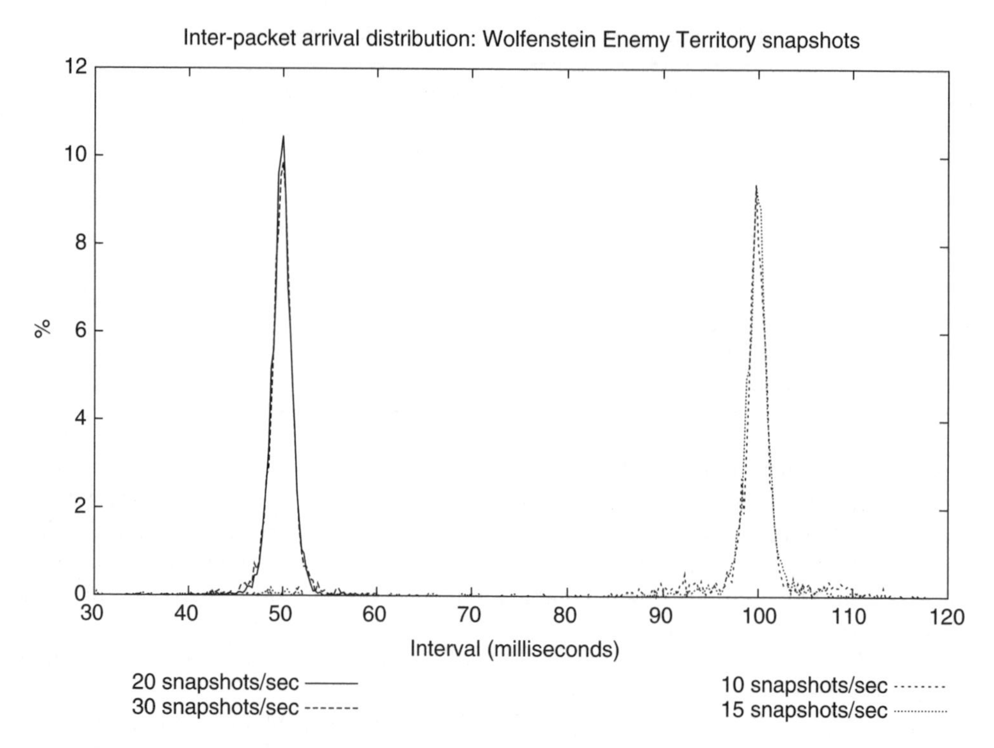

Figure 10.14 Inter-packet intervals for Wolfenstein ET snapshots towards a single client

10.4.1 Example: Wolfenstein Enemy Territory Snapshots

Figure 10.14 shows the inter-packet intervals for snapshot traffic from a Wolfenstein ET server to a single client. The client was approximately 12 hops and 15–20 ms away from the server over a regular (consumer grade) broadband Internet connection. Measurements were taken at the client end. There are four histograms, corresponding to the client requesting 10, 15, 20 and 30 snapshots per second. We can clearly see that the server has quantised the snapshot rate to an integer multiple of the server's internal tick interval of 50 ms when 20 or 30 snapshots/second are requested, and 100 ms when 10 or 15 snapshots/second are requested.

Figure 10.15 shows the inter-packet intervals for command traffic from a Wolfenstein ET client to its server (measured at a server 15 hops from the client).

10.4.2 Example: Half-life 2 Snapshots and Client Commands

Figure 10.16 illustrates the difference between HL2's snapshot transmission strategy and that of ET. As with Figure 10.14, the client is 12 hops (roughly 15–20 ms) from the server, inter-packet intervals are being measured at the client and the four histograms correspond to the client requesting 30 and 50 snapshots per second. At 30 snapshots per second, the HL2 server emits a stream of snapshot packets having a mixture of 30 ms and 45 ms intervals. At 50 snapshots per second, the actual snapshot stream is a mix of

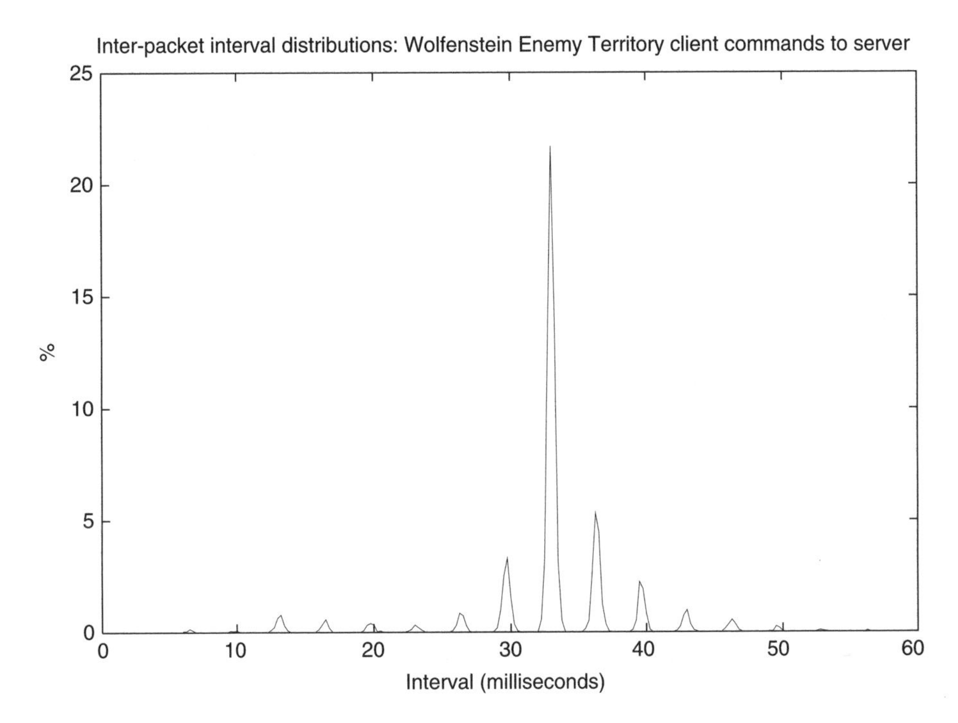

Figure 10.15 Inter-packet intervals for Wolfenstein ET command traffic from client to server

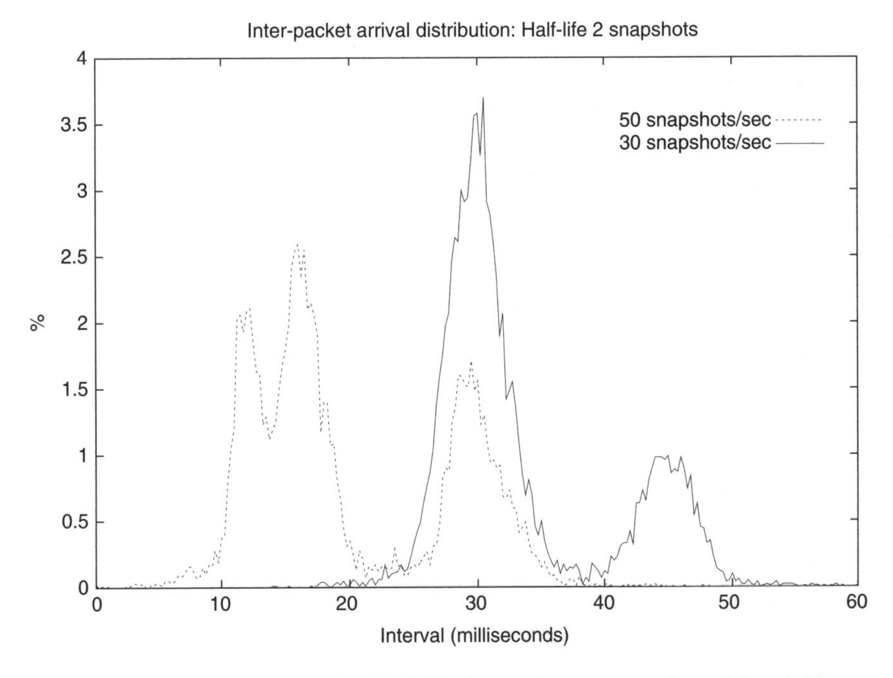

Figure 10.16 Inter-packet intervals for Half-life 2 snapshots to one client (30 and 50 snapshots per second)

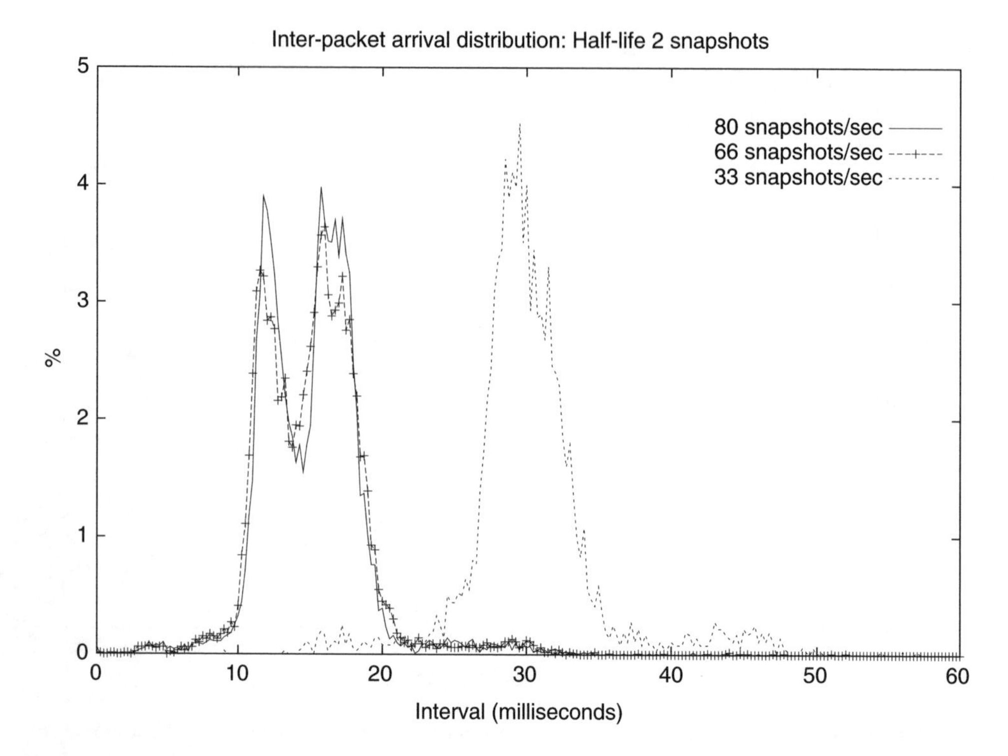

Figure 10.17 Inter-packet intervals for Half-life 2 snapshots to one client (33, 66 and 80 per second)

15 ms and 30 ms intervals. The average snapshot rate is what the client requested, but the actual inter-packet interval distributions are bi-modal.

HL2's inter-packet interval distributions for 33, 66 and 80 snapshots per second are shown in Figure 10.17. Since 33 snapshots per second is an even multiple of the tick timer, the distribution has a single peak at 30 ms. At 66 and 80 snapshots per second, we see identical distributions around 15 ms. Because of the way HL2 orders updates to individual clients per tick, the nominal 15 ms peak in the histogram appears as two peaks closely straddling 15 ms (~12 ms and ~18 ms).

Figure 10.18 shows the inter-packet interval distributions of client-to-server HL2 traffic as measured at a server. In this case, there were eight clients playing at the same time. We have plotted the inter-packet intervals of three representative clients and of the aggregate traffic from all eight clients as seen by the network link coming into the server. Two things are noteworthy.

- Each client has quite a distinct distribution of its own (because of client-side settings, CPU speed, graphics cards and player activity).
- The aggregate distribution is clustered well below the mean interval of each client (because client-to-server command traffic is not synchronised across clients).

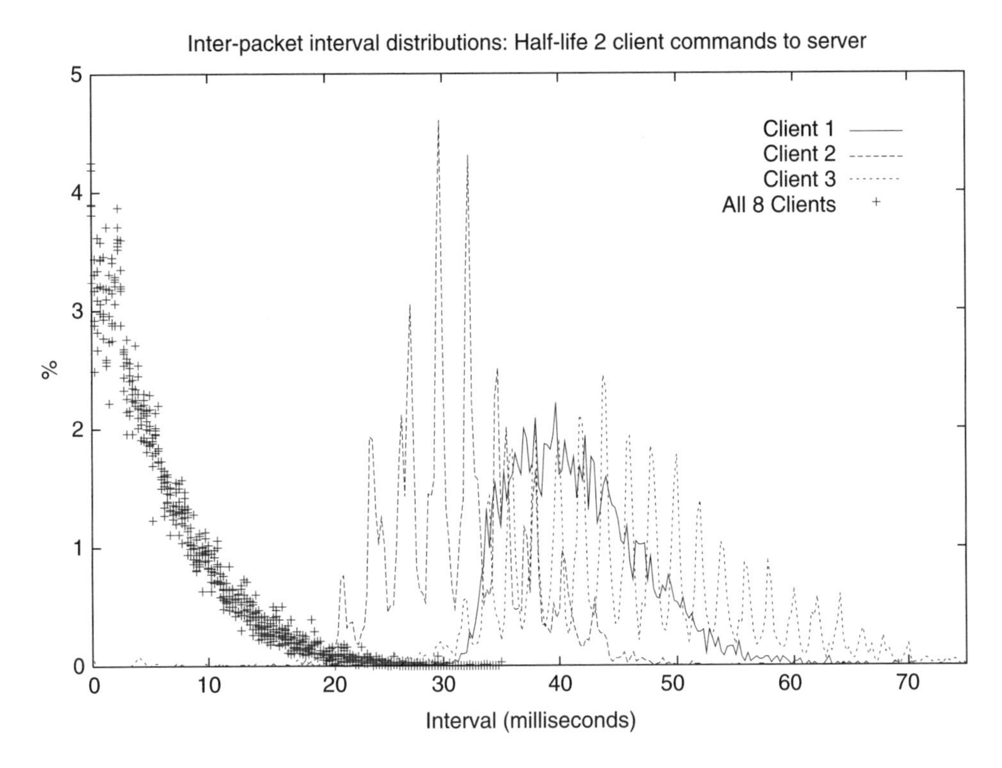

Figure 10.18 Inter-packet intervals for Half-life 2 client commands towards the server (three clients and aggregate of eight clients)

10.5 Estimating the Consequences

Many discussions of online games characterise game traffic in average 'packet per second' (pps) or 'bits per second' (bps) rates. Average packets per second is the inverse of the average inter-packet interval (for example, a 50 ms inter-packet interval equates to 20 packets per second). Bits per second represents the length of the average packet multiplied by the number of packets per second.

Unfortunately, the attractive simplicity of such single-value metrics hides the packet-by-packet realities revealed in packet-size distribution and inter-packet interval distributions.

For example, consider a Quake III Arena server with 15 players sending update packets every 50 ms to each client (default client settings). This server is transmitting 300 packets per second for an average inter-packet arrival time of 3.3 ms. If the average packet size was ~160- bytes (not unreasonable given Figure 10.7), that would equate to ~384 Kbit/second. However, as we know from direct traffic measurement, this stream is not uniform. Every 50 ms the server sends 15 packets back to back, as fast as possible – limited only by the server's hardware and local link speed.

Assume that link speed is the limiting factor, and the server is on a 100 Mbit/second ethernet connection. An average IP packet of 160 bytes translates to roughly 1472 bit-times on the ethernet [add 12 bytes for source and destination media access control (MAC) address, 2 bytes for ethernet type field, 4 bytes ethernet cyclic redundancy check (CRC) and 64 bit-times inter-frame preamble]. This translates to ~15 micro-seconds per IP packet, or 225 micro-seconds per burst. (Alternatively, this equates to a burst at roughly 67 K packets per second every 50 ms.)

The situation is substantially different if we consider the local server's link to be a 1.522 Mbit/second T1 link. For simplicity, let us assume 8 bytes overhead (e.g. for Point to Point Protocol (PPP), encapsulation [RFC1661]), so the average 160- byte packet becomes 168 bytes, or 1344 bits long. Each packet now takes 883 micro-seconds to transmit at 1.522 Mbit/second, and it takes 13.2 ms to send all 15 packets back-to-back. Thus, for 13.2 ms out of every 50 ms, the server transmits at 1136 packets per second.

These two examples show that the server's local link speed has a big impact on how bursty the server's outbound traffic will appear at early router hops along the path out towards the clients. In neither case does the server impose an even load of 300 packets per second on the network. In the latter case the regular update packets for the 1st and 15th clients are almost 12 ms apart.

Of course, the mean packet size can be quite misleading too. As Figure 10.7 suggests, depending on map type and number of players, a Quake III Arena server might easily find itself sending a burst of 300 byte packets at any given moment. Assume that at a particular instant the server update sends 15 packets of 350 bytes each. This translates to a worst-case instantaneous link bandwidth requirement of more like 840 Kbit/second – rather more than the nominal 384 Kbit/second calculated previously using average packet sizes. (The burst length would be 449 micro-seconds and 28 ms long at 100 Mbit/second and 1.522 Mbit/second respectively.)

The probability of such a worst-case burst happening depends on correlation (or lack of it) between the packet-size distributions of each server-to-client stream. Establishing this knowledge requires the kind of detailed traffic measurement discussed in section 10.1. Nevertheless, we can see that, for example, a 15-player Quake III Arena server could not reliably send client update packets with predictable latency if the local link was under 840 Kbit/second. (If you had provisioned a 512 Kbit/second link instead, there would be occasions when the Quake III Arena server would be unable to transmit all 15 update packets within a 50 ms update interval. Most likely this would cause update packets to briefly queue up inside the operating system of the server's host machine, adding temporary jitter to some of the server-to-client packet streams.)

10.6 Simulating Game Traffic

When an ISP wishes to explore the consequences of new network designs or deployment of new applications, it is generally impractical to simply build the new network and test it. The alternative is to simulate the interactions between new network configurations and new types of networked applications. Two questions arise from measurement of actual game traffic. First, can we create simulated traffic generators with similar statistical properties. Second, can we reasonably extrapolate from small numbers of clients to simulate the consequences of large client populations.

At least for a number of FPS games, it is not difficult to construct reasonably accurate statistical traffic generator functions. Some of the earliest work analysed Doom

[BORELLA99] without quite taking it to the level of simulation tool traffic generators. Subsequent work ([LANG2003, LANG2004] and [ZANDER2005a]) developed models for Half-life, Quake III Arena and Halo 2 respectively that could be used to specify ns-2 [NS2SIM] traffic generators.

Our aim is to approximate empirically observed traffic characteristics with a pre-existing statistical model that is both 'good enough' and simple to implement. (All other things being equal, simple traffic generator functions increase the speed with which simulations run and the scale of network topologies that can be simulated.)

10.6.1 Examples from Halo 2 and Quake III Arena

As an example, Halo 2's client-to-server inter-packet arrival times can be modelled as a normal distribution, while the server-to-client times were better approximated by an extreme distribution [ZANDER2005a]. Figure 10.19 shows the relationship between measured and synthesised intervals using the normal distribution for client-to-server traffic. (QQ-plots in [ZANDER2005a] show that errors in the synthesised distributions become noticeable only for 2 % of client-to-server traffic in the tails and 5 % of server-to-client traffic in the tails.)

Extreme distributions create a good match to packet sizes in each direction (Figure 10.20 and Figure 10.21).

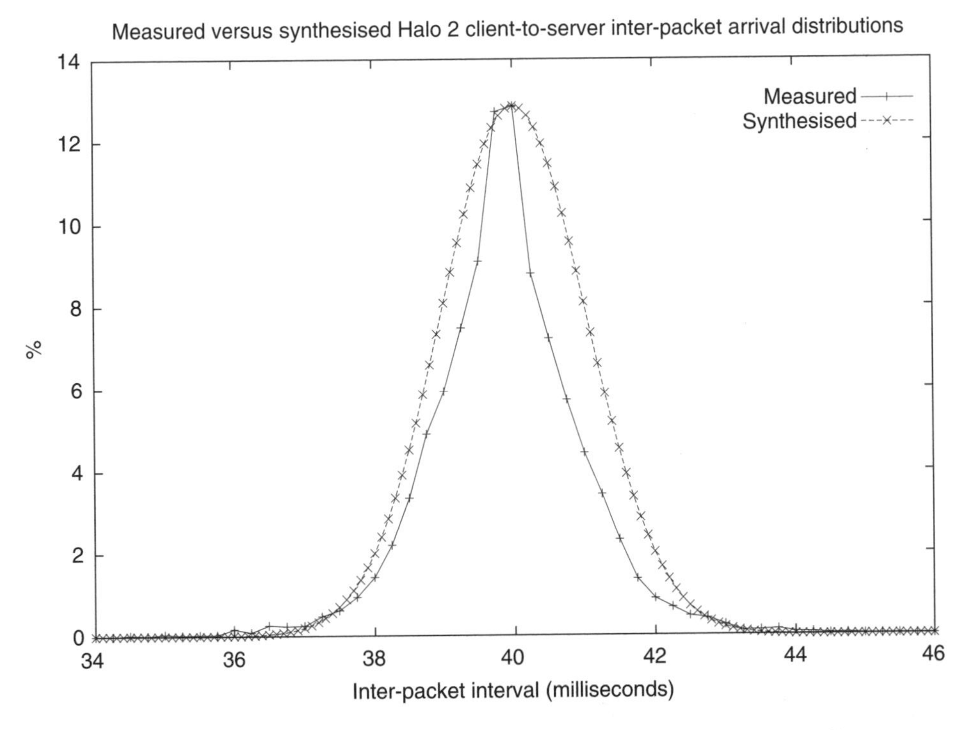

Figure 10.19 Measured and synthesised Halo 2 inter-packet arrival times for client-to-server traffic

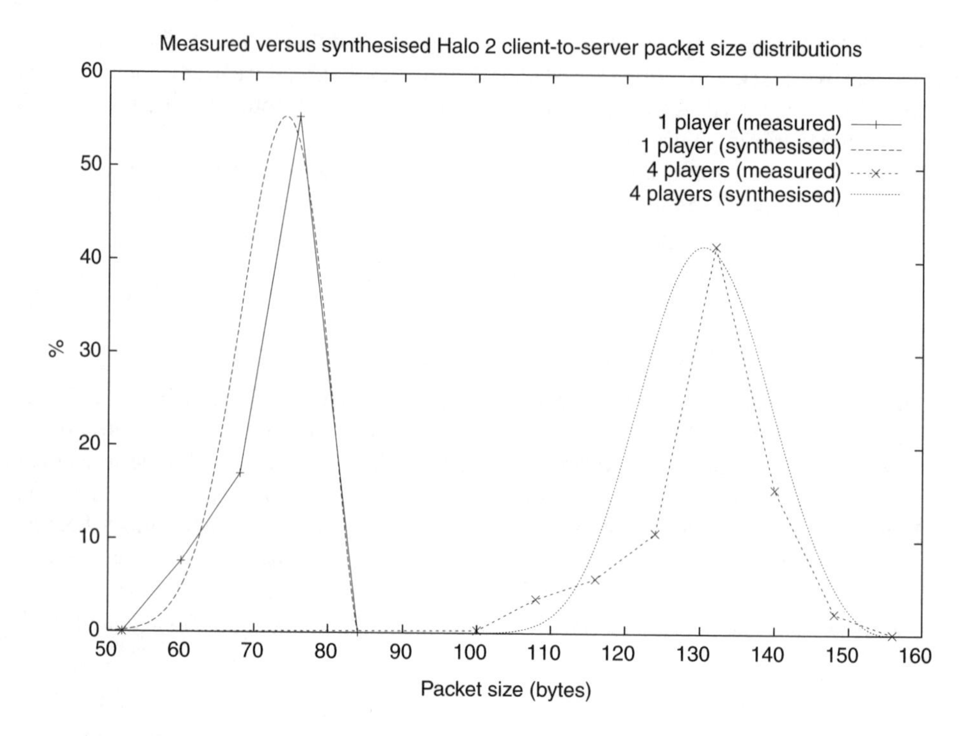

Figure 10.20 Measured and synthesised Halo 2 packet sizes for client-to-server traffic

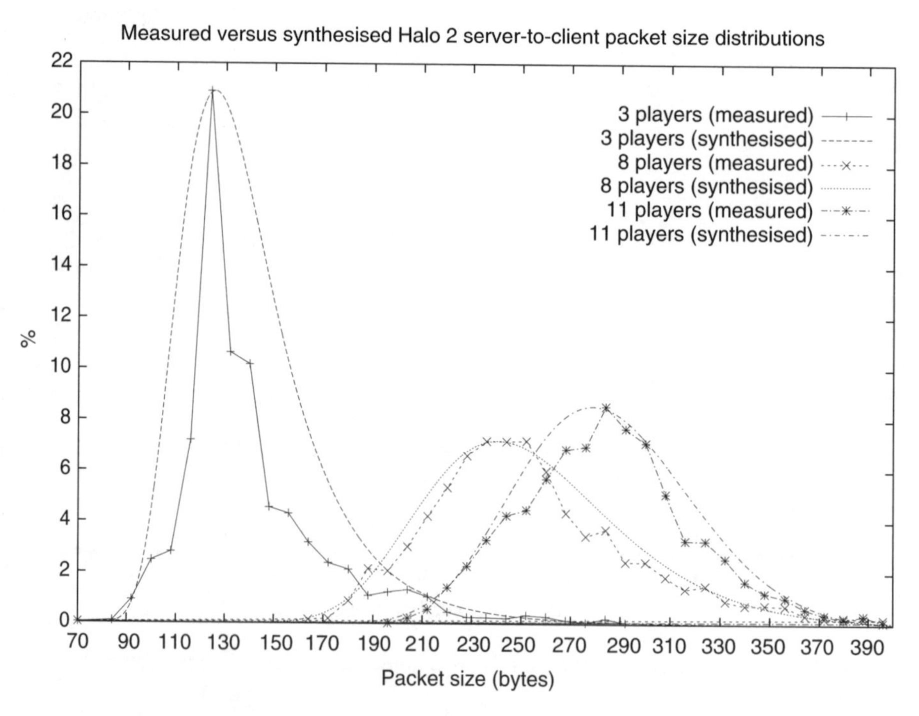

Figure 10.21 Measured and synthesised Halo 2 packet sizes for server-to-client traffic

```
// client to server traffic model
setMillisecondTimer(Random::normal(40,1))
 ...
timerExpired() {
   case numberOfPlayers == 1 :
     pktSize = Random::extreme(71.2, 5.7)
   case numberOfPlayers == 2 :
     pktSize = Random::extreme(86.9, 5.1)
          ...
   // round to nearest 8 byte packet size
    pktSize = round((pktSize-52)/8)*8 + 52
    sendPkt(pktSize)
 }
```

Figure 10.22 ns-2 code fragment for generating Halo 2 client-to-server traffic

An ns-2 code fragment to generate client-to-server packet traffic would look something like Figure 10.22 [ZANDER2005a]. (In this case the code utilises extreme distributions with different parameters depending on the number of players on the simulated client. The ns-2 code also mimics Halo 2's generation of packet lengths rounded to specific multiples of eight.)

A similar example is the analysis of Quake III Arena traffic performed in [LANG2004]. The authors ascertained that server-to-client packet-size distribution could be modelled as a lognormal ('... with mean 79.340,543 and standard deviation 0.24,507,092') for a two-client game, plus one additional exponential (with mean 13) for every additional client.

The code fragment in Figure 10.23 shows how easy this is to incorporate into ns-2. Here we simplistically assume that the packet generation occurs precisely every 50 ms (timer interval) and focus on the random variation in packet size depending on the number of clients. We also assume that every client receives a packet of the same size in each update interval. A more sophisticated ns-2 traffic generator model would randomise the packet size for each client, and slightly randomise the inter-packet intervals to match observed jitter in server-to-client transmissions.

It is not possible to itemise every possible traffic generator. What we should take from the preceding section is the knowledge that it is quite easy to simulate new network scenarios with relatively accurate models of game traffic.

```
/* the packet size is dependent on the number of players */
/* it is the base packet size distribution (for 2 players) */
/* plus a negative exponential with mean 13 for every additional player */
size_ = int (Random::lognormal(79.340543, 0.24507092));
for (int i=3; i<=nrOfPlayers; i++)
  size_ += int (Random::exponential(13));
  /* send one packet to each player */
  for (int i=1; i<=nrOfPlayers; i++)
      send(size_);
 /* schedule the next transmission */
 /* interval_ is 0.05 sec */
timer_.resched(interval_);
```

Figure 10.23 ns-2 code fragment for generating Quake III Arena server-to-client traffic

10.6.2 Extrapolating from Measurements with Few Clients

Recent analysis of FPS traffic models suggests that we can plausibly synthesise traffic with many clients from the models developed by measuring traffic for only a small number of clients [BRANCH2005].

Given n clients connected to the server, each server-to-client snapshot packet carries state-change information derived from all n client-to-server command messages received in the previous snapshot interval (roughly speaking). Thus, the size distribution of the server-to-client snapshot packets can be approximated by the convolution of n client-to-server packet-size distributions plus a small negative exponential function. Analysis in [BRANCH2005] showed that, starting with a simple client-to-server packet-distribution model we could reasonably predict the server-to-client packet-size distributions actually measured for four, six and eight-player games of Quake III Arena. This principle is expected to hold for higher numbers of players and similar FPS games (since the logical extrapolation makes very limited assumptions about the underlying game engine design).

References

[BELLOVIN2002] S. Bellovin, "A Technique for Counting NATted Hosts", Proceedings of Second Internet Measurement Workshop, November 2002.

[RFC1661] W. Simpson (editor), "The Point-to-Point Protocol (PPP)," STD 51, RFC 1661, July 1994.

[BORELLA99] M.S. Borella, "Source models of network game traffic", Proceedings of networld+interop '99, Las Vegas, NV, May 1999.

[BRANCH2005] P. Branch and G. Armitage, "Towards a General Model of First Person Shooter Game Traffic", CAIA Technical Report 050928A, Centre for Advanced Internet Architectures, Swinburne University of Technology, Australia, September 2005. (http://caia.swin.edu.au/reports/050928A/CAIA-TR-050928A.pdf)

[ETHEREAL] Ethereal, "Ethereal: A Network Protocol Analyzer", http://www.ethereal.com/ as of July 2005.

[FREEBSD] FreeBSD, "FreeBSD: The Power to Serve", The FreeBSD Project, http://www.freebsd.org as of July 2005.

[HALFLIFE] Planet Half-Life, http://www.planethalflife.com/half-life/, Accessed 2006.

[HALFLIFE2004] H A L F – L I F E 2, http://half-life2.com, 2004.

[HALO2004] Bungie.net, "Bungie.net : Games : Halo 2", http://www.bungie.net/Games/Halo2/, 2004.

[HL2NTWK] "Source Multiplayer Networking", http://developer.valvesoftware.com/wiki/Source_Multiplayer_Networking, as of August 2005.

[LANG2003] T. Lang, G. Armitage, P. Branch and H-Y. Choo, "A Synthetic Traffic Model for Half Life", Australian Telecommunications Networks & Applications Conference 2003, (ATNAC 2003), Melbourne, Australia, December 2003.

[LANG2004] T. Lang, P. Branch and G. Armitage, "A Synthetic Traffic Model for Quake 3", ACM SIGCHI ACE2004 conference, Singapore, June 2004.

[LINUX] Linux Online Inc, "Linux Online!", http://www.linux.org as of July 2005.

[NS2SIM] "The Network Simulator–ns–2", http://www.isi.edu.nsnam/ns/ (as of January 2006).

[QUAKE3] id Software, "id Software: Quake III Arena," http://www.idsoftware.com/games/quake/quake3-arena/, Accessed 2006.

[TCPDUMP] "tcpdump/libpcap," http://www.tcpdump.org as of July 2005.

[WET2005] "Wolfenstein," http://games.activision.com/games/wolfenstein as of October 2005.

[WINDOWS] Microsoft, http://www.microsoft.com as of July 2005.
[ZANDER2005a] S. Zander and G. Armitage, "A Traffic Model for the XBOX Game Halo 2", 15th ACM International Workshop on Network and Operating System Support for Digital Audio and Video (NOSSDAV 2005), Washington, USA, June 2005.

11

Future Directions

The future of online games holds lots of changes. Emerging technologies will affect network services in support of interactive, multiplayer games, new architectures will support massively multiplayer games in a variety of genres, server selection will become easier even as games crack down on cheaters, and novel game design will broaden the scope of online interactions.

11.1 Untethered

End-hosts will be increasingly untethered by wired networks. Online games of the future will travel over wireless networks to the end-clients. Wireless networks, because of their low cost and convenience of installation, are already becoming increasingly widespread. There are universities and even entire towns that have established wireless local area network (WLAN) access. Large cities are even in the process of planning ubiquitous, free wireless access for everyone.

Mobile telephone networks increasingly carry general, Internet Protocol (IP)-based traffic. These mobile networks have customers using increasingly sophisticated, yet low cost, mobile phones and personal digital assistants that are capable of playing a variety of interactive games. In fact, these mobile phones can have more computing power than early personal computers (PCs) that played games.

Gaming platforms are following suit. As mentioned in Chapter 3, the newest handheld game consoles are all enabled with Institute of Electrical and Electronic Engineers (IEEEs) 802.11 WLAN. The next generation of game consoles also follow this 'untethered' trend. Sony's Playstation 3 comes with a built-in Ethernet adapter, but also includes an IEEE 802.11 WLAN wireless adapter. The Xbox 360 comes with a built-in Ethernet adapter, but with an 802.11 wireless attachment purchased separately.

Figure 11.1 depicts the past, present and future of network connections for online games. Yesterday's online gaming networks were only wired, and with somewhat limited connectivity at that. Today's gaming networks are mostly wired, but with increased capacity to the end-user. More importantly, increasingly the last-mile connection is wireless. Beyond PCs and terminals, there is an increasing diversity of game devices that can network for online play. Today, it is the PCs, consoles and some limited mobile devices. Tomorrow, this trend will continue, with the variety of gaming devices growing, with most

Networking and Online Games: Understanding and Engineering Multiplayer Internet Games
Grenville Armitage, Mark Claypool, Philip Branch

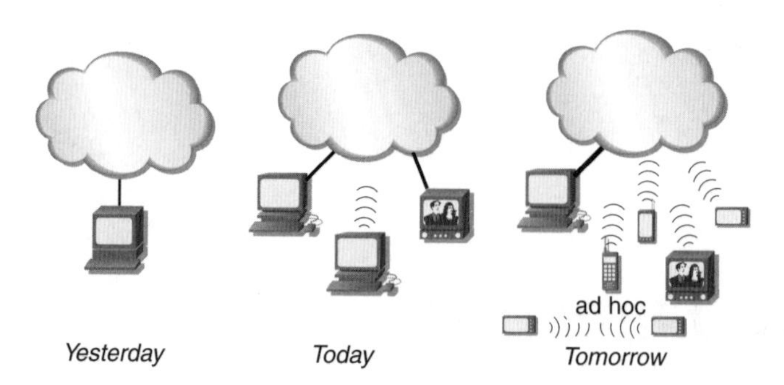

Figure 11.1 Network connections for online games yesterday, today and tomorrow

devices connecting via a wireless connection, although high-capacity wired connections will still continue.

This pervasiveness of wireless will impact online games considerably. Even for cases where a game server is on a well-connected, wired host, the game clients will often be wireless, through a WLAN or a wireless wide area network. Moreover, *ad hoc* wireless networks, where hosts do not join a fixed, preset network environment but rather form a network with willing hosts in range, are a natural mechanism to support online games for players in close proximity. These *ad hoc* networks present additional challenges of routing, stability and security all of which will impact the performance of game traffic they carry.

11.1.1 Characteristics of Wireless Media

The characteristics of wireless networks that have the most impact on online games and that differ from wired network are [PK02]:

(a) *Shared Medium.* Unlike the wired media of today, broadcasting is natural in wireless networks since all transmissions share the same medium. However, this shared medium causes collisions as computers transmit at the same time, and can suffer from interference even from computers transmitting on different channels. This can degrade network performance beyond today's Internet congestion that commonly occurs at routers or at the network interface cards. Finally, the shared medium also makes it impossible to increase the total capacity simply by adding media, as can be done in a wired network. With the wireless medium, the network is restricted to a limited available band for operation, and cannot obtain new bands or duplicate the medium to accommodate more capacity.

(b) *Propagation.* Wireless radio transmissions propagate over the air and are suspect to attenuation, reflection, diffraction and scattering effects. The multipath fading caused by these effects results in time-varying conditions, meaning the received signal power (and hence, the network performance) varies as a function of time. In today's wired networks, the physical performance on a given link is fairly consistent over time.

(c) *Large, bursty channel errors.* Owing to the attenuation, interference, and fading effects, wireless networks suffer from higher loss rates than typical wired networks. Wireless networks can see bit error rates of 10^{-3} or even higher and often in bursts. To accommodate these errors, most of the wireless network standards implement a variety of error recovery mechanisms, such as the Forward Error Correction (FEC), Automatic ReQuest (ARQ) for retransmission and rate adaptation.

(d) *Location dependent carrier sensing.* In wireless networks, performance is greatly determined by location. A hidden terminal where one sender is inside the range of the intended destination but outside of range of the sender causes collisions if the sender and the hidden terminal transmit to the destination at the same time. Similarly, an exposed terminal where one node is within the range of the sender but out of interference range of the destination may cause unexpected backoffs when an exposed terminal is transmitting, even if that transmission will not collide with the sender's transmission at the destination.

The spread of wireless networks that will support tomorrow's online, network games demand increasing attention be paid to the above effects on games.

11.1.2 Wireless Network Categorisation

The general way to categorise wireless networks is based on the range of coverage.

(a) *Wireless Personal Area Networks* (WPANs) operate within a confined space, such as a small office workspace or single room within a home with a coverage of less than 30 feet. For example, Bluetooth, which is defined under IEEE 802.15.1, can provide up to 720 Kbps capacity over less than 30 feet distance. Ultra Wideband (UWB), defined in IEEE 802.15.3a (still under development) is designed to provide up to 480 Mbps throughput over a short distance [Intel04].

(b) *Wireless Local Area Networks* (WLANs) have a broader range than WPANs, but are still typically confined within a single building such as a restaurant, store or home. WLAN has become perhaps the most popular wireless data communication technique with the production of the IEEE 802.11 standards.

(c) *Wireless Metropolitan Area Networks* (WMANs) cover a much greater distance than WLANs, connecting buildings to one another over a broader geographic area. For example, the emerging WiMAX technology, IEEE 802.16d today and IEEE 802.16e in the near future, will further enable mobility and reduce reliance on wired connections. Typical WMANs have a throughput of up to 10–20 Mbps and can cover a distance of approximately several miles [Intel04].

(d) *Wireless Wide Area Networks* (WWANs) have the broadest range and are most widely deployed today in the mobile voice market, where the WWAN provides the capability of transmitting data. The most popular WWAN techniques include the cellular 2.5 G data services, such as General Packet Radio Service (GPRS) and Enhanced Data Rates for GSM Evolution (EDGE), and the next-generation cellular services based on various 3G technologies.

The following table summarises approximate characteristics for the above networks.

Technology	Use	Range	Capacity
WPAN	Single room	100 m	1 Mbps
WLAN	Single building	1 km	50 Mbps
WMAN	Multiple buildings	10 km	20 Mbps
WWANs	Entire country	1000 km	100 Kbps

Out of these wireless network techniques, WLANs and WWANs are the most widely deployed wireless networks. WLANs are already used for all sorts of data communications over PCs, including online games, but the use of WWANs for online gaming is just beginning. The next generation of WWAN networks promise to provide even higher capacities than the current standards, but latency problems (300 ms to 1000 ms) are still formidable.

11.2 Quality of Service

Quality of Service (QoS) has been a popular, if challenging, research area for over a decade (Figure 11.2). While there have been a number of standards and approaches published, most notably Integrated Services (IntServ) [BCS94], the Resource Reservation Protocol (RSVP) [BZB+97] and Differentiated Services (DiffServ) [BBC+98], the deployment today of these technologies has been lacking.

Approaches that try to provide some QoS without signalling rely upon packet inspection to classify traffic into QoS classes. This means the IP header and the transport [transmission control protocol (TCP)/User Datagram Protocol (UDP)] header (typically, to get port numbers) and sometimes the payload are used to infer what type of traffic the flow is carrying. However, inspecting packet contents (transport and application layer) is becoming increasingly problematic with encrypted data and even 'well-known' port numbers are often not a reliable means of classifying traffic. The application-layer protocols themselves can be very complex, and often change making the QoS devices themselves very brittle.

Yesterday's networks were not well-connected or provisioned, depicted in the picture on the left of figure 11.2 by the narrow network 'pipe', and carried mostly one class of traffic as depicted by the gray squares. Today's networks, and end-host operating systems, are increasingly well-connected and provisioned, as represented by the thicker network pipe in the picture in the middle. However, today's networks still almost exclusively use 'best effort' services that do not provide explicit QoS guarantees, hence all the squares are

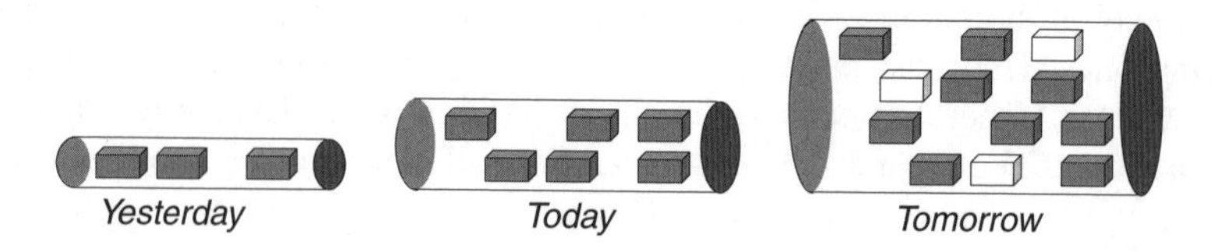

Figure 11.2 Quality of Service (QoS) yesterday, today and tomorrow

still gray. Tomorrow's networks will be of even higher capacity with the biggest network pipes and will provide opportunities to classify traffic based on some QoS requirements. This is indicated in the picture on the right by the classification of some packets as white, amongst the gray squares.

QoS can be characterised as either being parameter-based or priority-based. Parameter-based QoS specifies a strict QoS such as latency bound or required data rate [BCS94]. Priority-based QoS specifies relative importance, such as this application's data is more important than that one [BBC+98]. In general, per-flow, parameter-based QoS on the Internet has been of interest, yet is difficult to achieve end-to-end because of signalling and scalability concerns. Priority-based QoS can be more easily deployed on select nodes in a network.

11.2.1 QoS and IEEE 802.11

Wireless LAN equipment is low cost, convenient to deploy and use, can provide high user bit rates (typically in the Megabits per second range) and, provided care is taken in limiting sources of interference, is reliable. However, as mentioned above, wireless links can have high loss rates, bursts of lost frames, and high latency and jitter, making QoS difficult. And delay-sensitive applications, such as online games, require QoS guarantees, such as bandwidth, delay and jitter. Providing QoS for upper layer applications is one of the most challenging functions the wireless media access control (MAC) layer is trying support.

To meet this challenge, the dominant WLAN protocol, IEEE 802.11, has one incarnation, the 802.11e standard, designed to provide some QoS performance. As described in Chapter 8, IEEE 802.11e provides priority-based QoS to 802.11 to allow different kinds of applications (e.g., file download, voice over IP and online games) to be placed in different transmission queues with its own contention parameters. This allows queue bounds, backoffs during congestion and other channel characteristics to be managed on an application class basis. With these developments, 802.11e will be able to provide QoS guarantees. It is in many ways one of the most promising and significant developments in broadband access networks in recent years.

11.2.2 QoS Identification

One of the difficulties in QoS approaches is the requirement for applications, such as the games themselves, to indicate their explicit QoS requirements to the underlying network. Requiring this assumes developers are aware of QoS issues and understand the signalling parameters and its ramifications for their applications. But QoS requirements for games are still not well-understood and are hard to state explicitly.

Since backbone Internet network links are often overprovisioned and thus have little queuing delay or variance in delay, one approach is to provide QoS across the 'last-mile' link between the Internet Service Provider (ISP) and the end-host. One such approach is the Traffic Classification and Prioritization System (TCAPS) [SAB+05], depicted in Figure 11.3. TCAPS includes a traffic classifier that identifies traffic on the basis of its characteristics and breaks it into two groups: real-time/interactive (i.e., online games) and everything else. Flows identified as real-time/interactive are given priority over all other traffic travelling via the customer/ISP link. The classification method of TCAPS

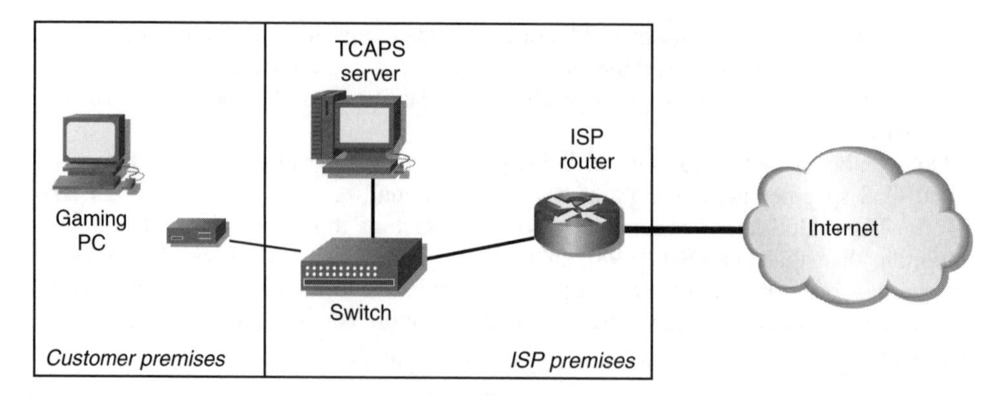

Figure 11.3 The Traffic Classification and Prioritization System (TCAPS) architecture aims to provide classification, and then QoS, to end-host customers

is flexible, removing the need for regular updates with new classifier information and removes any signalling burden from the end-hosts.

Since packet inspection is slow and unreliable, machine learning approaches are applied to learn the traffic patterns of certain kinds of traffic, so that the classifier will be able to make decisions about multiple flows much more quickly with much less information and processing. Work on synthetic traffic models for traffic that needs QoS, such as online games, can be effective in helping this approach.

11.3 New Architectures

Online games of the future will exploit new architectures. While online game architectures today are primarily client–server, peer-to-peer architectures naturally fit the model of communication for many games. Especially in models where players know and trust each other and use local predictions of server state, a centralised server plays the role of a glorified message carrier, just taking messages from players and forwarding them to other players. The peer-to-peer architecture allows clients (the peers) to send messages directly to other clients, thus avoiding the need (and added overhead) of a server. This provides a more natural model for network communication and has the potential for scalability as it more readily removes a centralised, potentially bottlenecked, server. Bottlenecked servers that seek to host more and more players on the same world, will incorporate increasingly sophisticated back-end architectures.

Figure 11.4 depicts the online game architectures of yesterday, with a few computers connected to an even fewer number of game servers shown in the picture on the left. Today, there are numerous game servers, but online games still primarily deploy a client–server architecture shown in the picture in the middle. Today's servers, however, have sometimes become more than a single machine hosting all the game players but have more sophisticated back-ends for load balancing. Tomorrow, online games will use a mixture of peer-to-peer and client–server traffic, shown in the picture on the right, and servers will be increasingly sophisticated.

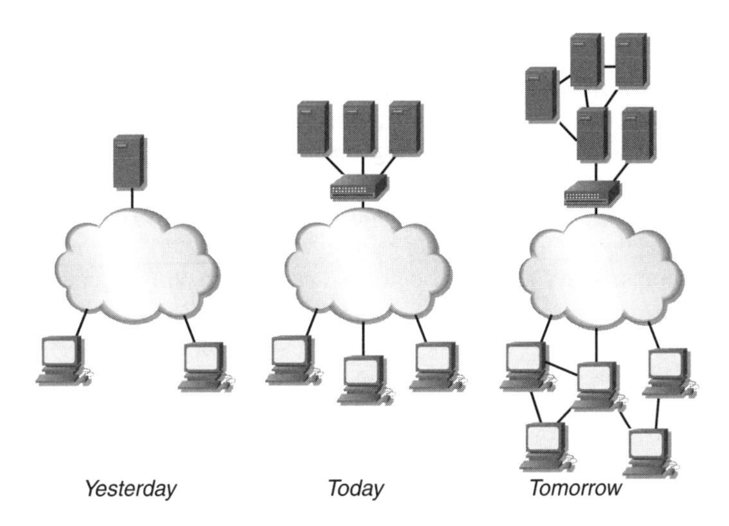

Figure 11.4 Online game architectures, yesterday, today and tomorrow

11.4 Cheaters Beware

There will be an increasing emphasis on cheat protection. The human nature that drives cheaters will remain. Namely, cheaters derive pleasure in vandalism in creating havoc in an online game (relatively few) and dominance in an attempt to gain advantage over other game players (most). The anonymity that is provided by network games removes many of the social aspects that prevent cheating, making it more pervasive in online games than it is in, say, multiplayer games on the same computer. Still, as games integrate technology that allows more socialisation, such as the integration of voice over IP with online games and even cameras that convey video information, the social constraints may reduce some forms of cheating.

In terms of technology, online games have moved from being completely vulnerable to even simple cheating exploits, to many of today's games requiring game clients to have some form of cheat protection software to participate in online play. New techniques will include network protocols that both detect and prevent cheating and even work in untrusted environments.

In practice, the fight against cheaters may never be won. An analogy may be the battle against viruses and worms. Approaches that catch viruses and worms typically react to past exploits, shoring up software defences to stop continued contamination. Virus writers react by coming up with new exploits or writing about more sophisticated viruses, thus continuing the cycle. Similarly, well-known cheating exploits can be accounted for, but new techniques are likely to be uncovered, even as game and network technologies grow.

Still, it can be expected that protection against simple cheating exploits, such as packet repetition and packet replay, will be provided in nearly every online game of the future, just as virus scanning software and firewalls are a part of nearly every modern computer today.

11.5 Augmented Reality

The game world will interact with the real world. Online games will increasingly incorporate aspects of the physical world into the gaming world. This will happen in subtle ways, such as new input devices allowing natural, physical input such as movements and gestures, to be incorporated into game play. More pronounced will be motion capture systems that incorporate physical actions into the game play. The physical world itself will become part of more online games, with computers augmenting the reality of real-world rooms, stairs and even outdoor spaces with virtual, computer-supported game play.

These new, augmented reality games bring new challenges for the aspects of network games. Specifically, the amount of input can dramatically increase with motion capture devices. Users playing in the real world are likely to be mobile, raising challenges for degraded wireless networks or even routing for wireline networks.

11.6 Massively Multiplayer

To date, most massively multiplayer online (MMO) games are of the role-playing variety, with massively multiplayer online role-playings (MMORPGs) having the lion's share of the massively multiplayer game market. In general, role-playing games allow a player to enhance an avatar through various game-related tasks. While some MMO games have proven popular and lucrative for some game companies, the game genre has been relatively narrow (Figure 11.5).

However, there is nothing exclusive about large games needing to be role playing. Consider the world itself as a massively multiplayer endeavour. Many human interactions happen on a large scale, such as military encounters during war, voting during elections, spectator events and even sporting events such as the Olympics, that involve hundreds and even thousands of people. Expect to find MMO games in other genres, such as real-time strategy, first person shooter, and racing games, as game and network technologies evolve to support more players.

Moreover, online games that allow players to easily change perspective, from an overhead, impartial view of the worldlike in a strategy game, to a third person perspective

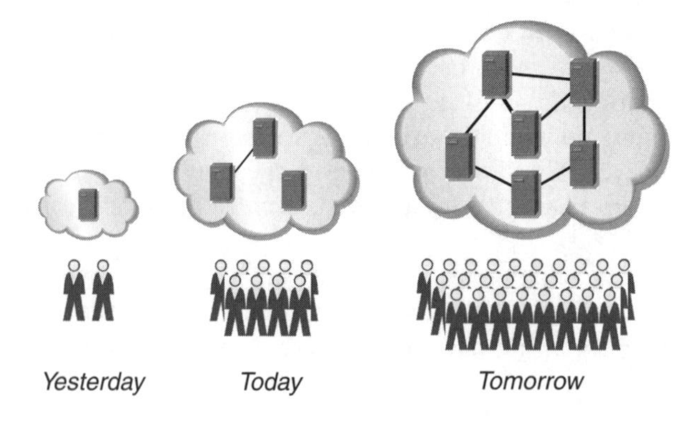

Figure 11.5 Multiplayer, online games yesterday, today and tomorrow

of particular avatars, to a first person view of an avatar. These perspective shifts each bring different network requirements with them, in terms of delay sensitivity and bit rate requirements, posing new challenges as the online game must adapt to changing user requirements with the changing perspectives.

11.7 Pickup and Putdown

Future online games will allow you to squeeze in a few minutes of game play whenever, and wherever, you can. Game design will emphasise game play that allows a pickup for a few minutes then a putdown. As technology becomes more pervasive in everyday life, interactive entertainment will follow. Expect an increase in the number of games on mobile devices, such as cell-phones and PDAs and for handheld game consoles to become more powerful and more portable. With it will come an increasing emphasis on games that allow players to pickup and putdown in a short amount of time. Such 'casual games' already exist on the Internet (and are amongst the most popular online games, in terms of number of players not network bandwidth) and allow online play for short breaks, but there will be an emphasis for such game style to permeate to games of interest to hard core game players.

11.8 Server Browsers

Many network games allow players to choose which server to connect to for their online play. For many games, this arises because individual users can run their own game servers, allowing clients to connect to their server from anywhere on the Internet. Nearly all popular first person shooters (such as Quake, Doom and Battlefield) allow individual users to run game servers. Similarly, most real-time strategy games (such as Warcraft and Age of Mythology) allow users to host a game, thus providing many choices for clients playing online.

And the choice of game server matters. Game servers can become full, limiting a player's ability to join the server. The requirement of some game servers for clients to have cheat protection enabled, or specific client versions installed may also physically limit a player's choices. The choice of the game map, game configuration and other in-game parameters (such as having friendly fire disabled for a team-based first person shooter) can determine a player's desire to join a particular game server.

Even if all physical and preferential game conditions are met by a game server, the network and server performance will impact the choice of the best server. The range of latencies from a client to all game servers can be as broad as the range of end-to-end Internet latencies. Moreover, game players care about application to application latencies, not just end-host to end-host latencies, so latency from server load adds to the network latency and makes selection of a close and fast server important for good online game play.

With current game server selection browsers, when a player starts a game client, the client contacts a master server that lists all game servers that are up at that time. The client then contacts each server individually to get latency information as well as server configuration parameters (map in play, number of players, etc.). A player can then sort the resulting list of game server information in a variety of ways, such as by increasing latency or by server map type. While it is somewhat flexible for

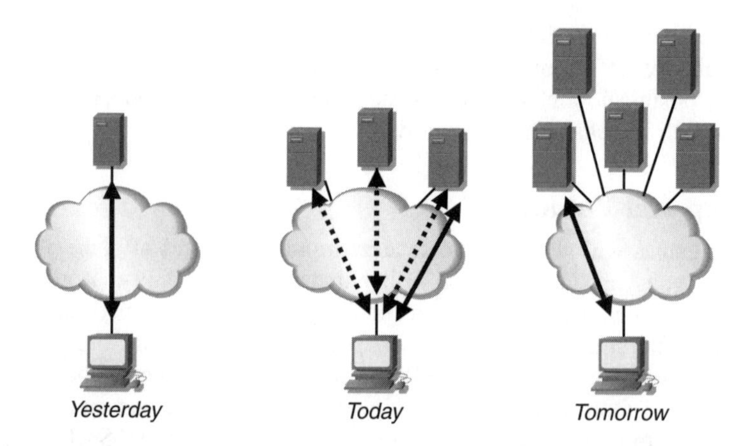

Figure 11.6 Server browsing yesterday, today and tomorrow

one player, current game server browsers provide little confidence on the performance of individual servers and no support for multiple players that want to play together (Figure 11.6).

The next generation of online game servers will make it easier to find 'good' servers, with some of this process even transparent to the user. There will be no need to contact every server in the world, sort through server lists, join a server only to find its performance lacking and then repeat the process. This will reduce the churn on game servers themselves, allowing their resources to be put to playing the game.

Yesterday, server browsing was easy (Figure 11.6). Online games had only one server they could connect to, depicted in the picture on the left. No browsing was needed as there were no choices. Today, many games have numerous servers from which to choose. Server browsing is primitive, often requiring players to connect to servers to find out how the game play is. This is depicted by the picture in the middle with the dashed lines showing connections and the solid line showing the server eventually chosen. Tomorrow, even though the number of server choices will increase, server browsing will be much more transparent, allowing quick connections to servers that perform well, indicated on the right by just the solid line.

References

[BBC+98] S. Blake, D. Black, M. Carlson, E. Davies, Z. Wang and W. Weiss, "An Architecture for Differentiated Services", IETF RFC 2475, December 1998.

[BCS94] R. Braden, D. Clark and S. Shenker, "Integrated Services in the Internet Architecture: an Overview", IETF RFC 1633, June 1994.

[BZB+97] R. Braden, L. Zhang, S. Berson, S. Herzog and S. Jamin, "Resource ReSerVation Protocol (RSVP) – Version 1", Functional Specification IETF RFC 2205, September 1997.

[Intel04] Intel Corporation, "White Paper: The New Era in Communications", [Online] http://www.intel.com/netcomms/bbw/302026.htm, 2004.

[PK02] K. Pahlavan and P. Krishnamurthy, "Principles of Wireless Networks – A Unified Approach", Prentice Hall PTR, 2002.

[SAB+05] L. Stewart, G. Armitage, P. Branch and S. Zander, "An Architecture for Automated Network Control of QoS over Consumer Broadband Links", *IEEE TENCON 05*, Melbourne, Australia, November 21–24, 2005.

12

Setting Up Online FPS Game Servers

To round out the book in this chapter, we will look at installing and starting dedicated servers for two specific first person shooter (FPS) games – Wolfenstein Enemy Territory (WET) and Half-Life 2. We will focus primarily on getting things started, and leave it to online forums and discussion groups to provide you with tutorials on specific aspects of running and maintaining a dedicated public server.

12.1 Considerations for an Online Game Server

Setting up a dedicated server raises many questions relating to expected performance, resource requirements and server monitoring. Given the number of players you wish to support at any one time you will need requirements estimates for CPU speed, memory (RAM), disk space, and network connection speed. It is usually advisable to monitor long-term server behaviour and usage patterns (as discussed in Chapter 9). This involves becoming intimately familiar with the server's own logging facilities and deploying network sniffing tools such as tcpdump. Sufficient diskspace must be allocated to hold server logs and packet tracefiles. Log rotation may be built into the game server (e.g. Half-Life 2) or you may need to implement your own logfile rotation scripts.

The network link capacity requirements can be estimated based on game server settings (such as snapshot and command transmission rates discussed in Chapter 10) and typical packet size distributions. (You will probably need to run some in-house trials of each game and map in order to measure the possible range of snapshot packet sizes.) If the server will be placed behind a NAT-enabled router, you will need to ensure the necessary ports are open on the router, so that people outside your network can properly connect to your game server.

Some FPS games allow a choice of server platform (usually Linux or Microsoft's Windows). In general, it is far better to use a dedicated PC running Linux or FreeBSD (with FreeBSD's Linux-compatibility mode enabled), because they will simply be more stable platforms than a Windows box. Aside from the zero cost of obtaining and installing Linux or FreeBSD, a Linux-based or FreeBSD-based server can also be managed remotely over the network. This makes it easier to hide the physical box in a cupboard, rack or a

Networking and Online Games: Understanding and Engineering Multiplayer Internet Games
Grenville Armitage, Mark Claypool, Philip Branch

hidden room. Not surprisingly your dedicated server does not require a high-end graphics card, or indeed any graphics card at all.

You may also wish to experiment with techniques for artificially balancing the latency experienced by different players. For example, in [LEE2005] a FreeBSD-based game server utilised dummynet [DUMMYNET][RIZ1997] (discussed in Chapter 7) to add small amounts of latency to each player's traffic – ensuring that all players 'experienced' much the same latency. (The game server in [LEE2005] would be regularly polled the server to monitor each player's experienced latency, adjusting the dummynet latency up or down as necessary.) Because latency can only be added, not removed, players can only be equalised if the target latency is higher than the latency usually experienced by your players. On the other hand, you want the target latency to still be 'low enough' (e.g. 100 ms) that people are still willing to play on your server. So, players experience latency equalisation only up to a point.

12.2 Wolfenstein Enemy Territory

Built on the Quake III Arena game engine technology, WET is a team-oriented FPS, based around a number of World War II-themed missions. Players join one of two teams ('Axis' or 'Allies') and take on a particular class of team members (Engineer, Soldier, Medic, Field Ops or Covert Operative). Although initially developed as a commercial add-on to Activision's 'Return to Castle Wolfenstein', the complete game software (both client and server) is now available for free download from Activision and a number of other sites [WET2005]. There is no need to purchase any CD, DVD or download license.

In this section, we will summarise the key steps in setting up an online WET server for multiplayer action. However, we will not provide a detailed discussion of WET itself from the perspective of game-play, player models, weapons, 3rd-party maps, etc.

12.2.1 Obtaining the Code

Wolfenstein Enemy Territory is available for Microsoft Windows, various Linux distributions and the Apple Macintosh. As of late 2005, the latest version of WET was version 2.60 (released on March 21, 2005). The releases are available from a number of sites on the Internet, including those shown in the list below. Generally these sites either allow direct, free download or will require a free registration before allowing downloads.

- http://www.3dgamers.com/games/wolfensteinet/downloads/
- http://www.planetwolfenstein.com/
- http://returntocastlewolfenstein.filefront.com/
- http://www.fileplanet.com/30846/0/section/Wolfenstein-Series

(If the above URLs do not work by the time you read this chapter, try searching from the main site in each URL or start back at the main Activision site, http://games.activision.com /games/wolfenstein.)

For version 2.60 you can download the Linux and Macintosh versions as a single installer. For Windows you download and install the full version 2.55, and then download

Table 12.1 Files required for WET version 2.60

Operating system and type	Filename	Release date	Size
Windows full version 2.55	wolfet.exe	28 May 2003	258 MB
Windows, patch to version 2.60	et_patch_2_60.exe	21 March 2005	5.5 MB
Linux full version 2.60	et-linux-2.60.x86.run	21 March 2005	258 MB
Macintosh full version 2.60	wolfet_2.60.dmg	21 March 2005	258 MB

(and run) the patch to version 2.60. These files are typically served from public sites using the file names listed in Table 12.1.

Each installer is an executable that, when run, prompts you for a location to place your WET files – the offered defaults are usually acceptable.

12.2.2 Installing the Linux Game Server

For now, we will assume that your intended server host has a keyboard and monitor and is running at least a basic text console (the installer uses relatively primitive text mode menus).

Create (or ask the system administrator to create) a dedicated user account for running the WET server (for example, user name 'et' and a home directory of '/home/et/'). This helps keep all your WET server activities distinct from other users on the server host. Wolfenstein Enemy Territory does not need root (administrator) privileges to run. For the rest of this description you are assumed to be logged in as the 'et' user.

Download et-linux-2.60.x86.run to your home directory, set the executable permission bits and launch the installer with:

```
/home/et/% chmod 555  ./et-linux-2.60.x86.run
/home/et/% ./et-linux-2.60.x86.run
```

The installer first unpacks the entire ET distribution (about 272 MB of files) before prompting you to accept a license and specify where the game should be installed. If your host has less than 300 MB of free space in /tmp (the usual temporary files directory), you should explicitly specify a different extraction directory as a parameter to the installer. For example, if you have plenty of space in your home directory you could use:

```
/home/et/% ./et-linux-2.60.x86.run --target /home/et/ettmp
```

A new directory, /home/et/ettmp, will be automatically created and used by the installer. However, you will need to manually delete this directory and its contents once the installation has run to completion.

Although the installer prompts with a default location of '/usr/local/games/enemy-territory/', we recommend placing the server's installation location under '/home/et/enemy-territory' (Figure 12.1). This ensures your WET installation is localised under the 'et' user's home directory (valuable if you do not have 'root' access on the server

Figure 12.1 Installer requesting installation location

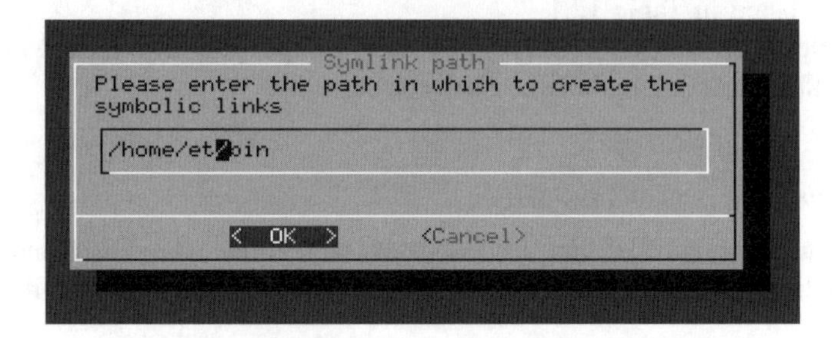

Figure 12.2 Installer requesting symlink location

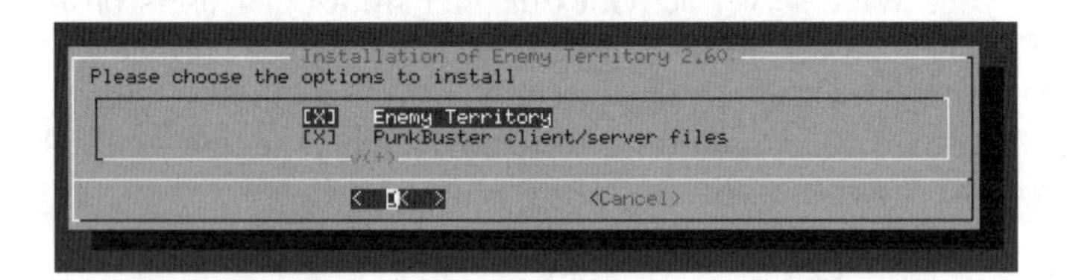

Figure 12.3 Chose both ET and PunkBuster

host, or you are running multiple game servers on a single host – separate instances of WET can be installed under different user accounts).

The installer will also ask for a location to place two symlinks for WET (Figure 12.2). We suggest /home/et/bin rather than the default /usr/local/bin (ensure /home/et/bin exists and is listed in your user's PATH environment variable before launching the installer). These symlinks allow you to launch the et client simply by typing 'et', and the et dedicated server simply by typing 'etded' at a command line prompt.

You then have the option of installing PunkBuster along with the main WET client and server code (Figure 12.3) [PBUSTER]. We recommend installing PunkBuster even if you do not expect to use it all the time. You will then be asked to accept the PunkBuster license conditions.

Files from the temporary directory (/home/et/ettmp, if you followed the example above) are then copied over to '/home/et/enemy-territory'. You are finally offered a choice of automatically creating KDE or GNOME desktop shortcuts before installation completes.

Ultimately, you will have 270 MB of files installed under '/home/et/enemy-territory' along with symlinks from '/home/et/bin/et' and '/home/et/bin/etded' pointing to the same named files under /home/et/enemy-territory/.

If you manually specified a directory to unpack the installation, you will now need to manually clean it up with:

```
/home/et/% rm -rf /home/et/ettmp
```

Because the WET installer uses a text-based interface you may also do this installation process while remotely logged into the server host from elsewhere (for example, over an ssh connection [SSH2005]). This is most convenient when your server host is already mounted in remote rackspace without a dedicated keyboard and monitor.

12.2.3 Starting the Server

Having installed WET you now have access to the official documentation on your local filesystem, located at:

```
/home/et/enemy-territory/Docs/Help/index.htm
```

You now have two choices – either launch a 'listen server' from the game client's graphical interface or run a dedicated server with no graphical interface. Our interest is solely in the latter case – you are running a dedicated server on a host that is tucked away in a rack in a closed room somewhere.

A couple of technical decisions need to be made before starting your dedicated server.

- What UDP port will it run on?
- Is this for local LAN play, or available for anyone on the Internet?
- What is the maximum number of supportable players?

(There are also some game-specific questions you need to consider, such as 'what game type and map rotation do I want?'. In part, these are subjective decisions that depend largely on your motives for running a server. Detailed descriptions of how to set up various game types are outside the scope of this book.)

By default, a WET server will bind itself to UDP port 27960 on the local server host (or automatically try one port higher if the default is in use). However, you can explicitly specify an alternative port if another process on the server host is already using port 27960 (for example, another WET server, or a Quake III Arena server).

You can specify whether or not your dedicated server registers with idSoftware's Enemy Territory master server (etmaster.idsoftware.com). You should not register if you just want to run a private server (even if it is on the Internet) or if you are running locally on an isolated LAN.

Player slots should be limited based on your available network capacity and processor speed. The official documentation (noted earlier) recommends '30 MHz per player slot', and thus recommends ~1 GHz for 32 players, 500 Mhz for 16 players and 250 MHz for eight players. Use the discussion in Chapter 10 to estimate your per-player network bandwidth requirements.

The usual way to start a dedicated server is to run 'etded' with a number of command line options. For example:

```
/home/et/% etded +set net_port 27985 +set dedicated 2 +exec
   server.cfg
```

This particular set of options instructs the server to run on a non-standard UDP port 27985 ('+set net_port 27985') and register with etmaster.idsoftware.com:27950 ('+set dedicated 2'), then pull the rest of its configuration details from a file named 'server.cfg' located in one of two locations:

- /home/et/enemy-territory/etmain/
- /home/et/.etwolf/etmain/ (or other, specified by fs_homepath)

In fact, almost everything you might want to configure can be placed in server.cfg. A default server.cfg is created when you first installed WET, which documents many useful parameters and settings. An alternative configuration file can be specified on the command line with:

```
/home/et/% etded +exec alternative.cfg
```

The file 'alternative.cfg' will be searched for in the two directories mentioned above, and used instead of the default server.cfg file.

If you are running on a host with multiple IP interfaces, you may wish to explicitly identify which IP interface should be advertised by the WET server. For example, the following command line specifies that the server advertises itself (and only 'exists' on) IP address 192.168.1.50:

```
/home/et/% etded +set net_ip 192.168.1.50 +exec server.cfg
```

(In this case, unless server.cfg contained any directives to the contrary, the server will attempt to use UDP port 27960.)

Add '+set sv_maxclients NN' to establish a limit of NN players on your server (the default is 20).

If you are simply experimenting while deciding what configuration parameters to use, note that the server does not actually 'go live' until it is told what map to play. If you have no map specified in your server.cfg file, use '+map MMM' on the command line (e.g. '+map oasis').

If you chose to run a public server, you will begin to see inbound probe traffic from remote clients within minutes (or even seconds) of your server registering with etmaster.idsoftware.com (as discussed in Chapter 9).

12.2.4 Starting a LAN Server

Use '+set dedicated 1' to establish a private, unadvertised server that will only respond to direct queries from people who already know the server's IP address and port number.

For example, use something like:

```
/home/et/% etded +set dedicated 1 +map oasis
```

This creates a minimalist, local server starting on the 'oasis' map which can be probed by, and connected to by clients on your LAN. (In the absence of other configurations, the server will enter the 'campaign' mode, and eventually move on to the next map in the campaign sequence.)

Clients can discover servers on their local LAN by broadcasting server-discovery 'get-info' probes (UDP packets sent to IP address 255.255.255.255). Each time the client presses the 'refresh list' button on their PlayOnline ServerBrowser server selection screen (when in 'Source:Local' mode), eight UDP broadcast probes are sent – two each to ports 27960, 27961, 27962 and 27963. This normally triggers a response from any of the dedicated servers on the local LAN that are listening on one of the four common WET ports.

Note one interesting caveat: if you run a LAN-only game using '+set dedicated 1' and also specify the game server's local IP address with '+set net_ip w.x.y.z' the server appears to ignore the server-discovery queries being sent to 255.255.255.255. In this case, each client will need to manually specify the server's IP address and port number in order to connect.

12.2.5 Ports You Need Open on Firewalls

It is entirely possible you have a firewall between your server and the outside world. If so, you must ensure that the server can send packets outbound to port 27950 on remote Internet hosts (to register with etmaster.idsoftware.com, and possibly other master servers) and receive inbound packets to whatever port you run the server itself (typically 27960).

12.2.6 Dealing with Network Address Translation

A number of additional considerations exist if your server host is behind a router doing Network Address Translation (NAT). This can occur if your server is on a home or small corporate LAN sharing a single broadband connection to the outside world. As noted in Chapter 4, NAT (or more precisely NAPT) makes multiple hosts in a hidden network appear to be a single, large host to the rest of the Internet. This is achieved by mapping all the hidden (or 'inside') IP addresses to a single external IP address, and then re-mapping TCP and UDP port numbers to ensure every application flow retains a unique 5-tuple of IP source and destination addresses and port numbers.

An issue arises when you try to run a public server from behind a NAT router. The server registration process involves your server sending the ASCII text 'heartbeat ETServer-1' to etmaster.idsoftware.com:27950 in a single UDP packet. (If your server subsequently shuts down gracefully, it will send the ASCII text 'heartbeat ETFlatline-1' in another UDP packet to etmaster.idsoftware.com:27950, removing your server from the master server's list of active servers.)

Upon receipt of the registration message, the master server engages in a short handshake to confirm your server is up and valid. Your server is now advertised as 'being at' the IP source address and source UDP port number from which your server's registration message arrived. If your server is behind a NAT router, the master server 'sees' the public IP address and port number used by the NAT router to represent your internally hosted game server.

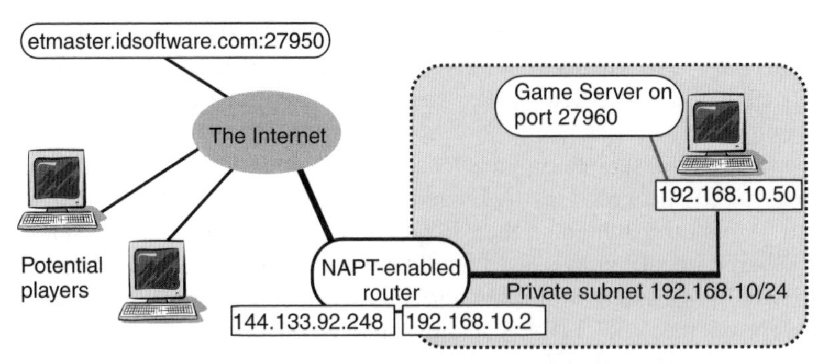

Figure 12.4 Example of game server behind a NAT router

For example, let us assume that your server is on a local LAN at 192.168.10.50:27960 and your NAT router's external, public IP address is 144.133.92.248 (Figure 12.4). Your game server's registration packet was re-mapped on the way to etmaster.idsoftware.com so that it appeared to originate from 144.133.92.248:47831. Your NAT router simultaneously established another internal mapping such that UDP packets coming back from etmaster.idsoftware.com:27950 to 144.133.92.248:47831 are re-mapped and forwarded on the internal LAN to 192.168.10.50:27960.

The problem is that your NAT router has no rules to handle UDP packets coming in to 144.133.92.248:47831 from anyone else on the Internet. Prospective players probing your server 'at' 144.133.92.248:47831 will receive no answer.

The solution is to manually configure your NAT router such that all packets coming in to 144.133.92.248:47831 from anywhere are re-mapped and forwarded on the internal LAN to 192.168.10.50:27960. In fact, the general approach is to create a bi-directional mapping rule stating (effectively) that '144.133.92.248:47831 on the outside is equivalent to 192.168.10.50:27960 on the inside network, regardless of who wants to talk to us'. (Every NAT-capable router has its own technique and terminology for configuring such mappings. You will need to consult the userguide for your own router for details.) With this additional rule, your game server will be playable by people located on the public side of your NAT router.

You may use any free UDP port you like on the public side of your NAT router. Players who find you through the automated server-discovery mechanisms will usually be unaware of your server's actual port number – so long as they can probe your server and connect they will be happy.

12.2.7 Monitoring and Administration

When started from the command line the game server prints information to the screen and accepts keyboard input in real time, allowing you to monitor server state and modify server configuration parameters. For example, you could type 'status' to find out the names of all clients currently connected, along with their current ping time, IP address and port number.

However, you will usually want to start the game server and then leave – returning only occasionally to monitor activities and perhaps modify parameters. For this you need to enable logging and 'rcon' (remote console).

To track server activity, you enable logging with '+set g_log ⟨logfilename⟩' – either on the command line or in the server.cfg file you use, to carry the server's main configuration settings. The server log output will then be directed to the file ⟨logfilename⟩. By default, this file will be placed under '/home/et/.etwolf/etmain', but the actual location can be changed with '+set fs_homepath ⟨newdirectory⟩'. So, for example, the following line:

```
/home/et/% etded +set g_log mylog +set fs_homepath /home/et/blah/
```

would force the game server's log output to be sent to '/home/et/blah/etmain/mylog'.

Changing fs_homepath is useful when running multiple instances of WET server on the same host – you can use a separate fs_homepath for each server. The main WET server, mods and PunkBuster will create their logging and configuration subdirectories below fs_homepath.

A relative fs_homepath is interpreted relative to '/home/et/enemy-territory/'. Thus using:

```
+set fs_homepath blah
```

the logfile is /home/et/enemy-territory/blah/etmain/mylog, and using:

```
+set fs_homepath /home/et/blah
```

the logfile is /home/et/blah/etmain/mylog.

If you are using a modified game (a mod, such as ETPro [ETPRO05]) the logfile will appear under the mod's name inside the directory pointed to by fs_homepath, rather than under 'etmain'.

Remote control of your game server is possible through the 'rcon' facility. Rcon is a client-side console command that passes text to the server, and prints the server's output back onto the client's console window. Rcon access to a server is enabled by setting an rcon password with '+set rconpassword ⟨password⟩'.

For example, to set the rcon password to 'mypass' you would start the server with:

```
+set rconpassword mypass
```

Then from a client you could execute the 'status' command on the server with:

```
/rconpassword mypass
/rcon status
```

(You only need to enter the /rconpassword line once, then use multiple/rcon commands as necessary.)

Rcon functionality is also available in tools such as the GUI-based XQF server-discovery tool [XQF05]. Figure 12.5 shows XQF's Rcon console in action – in this case the number of players has been checked, the server's name checked, then changed from 'myserver' to 'newservername' and then the change confirmed.

Unfortunately, the rcon mechanism has an important weakness. Every time an rcon command is executed, the client (or third-party application such as xqf) sends the rconpassword

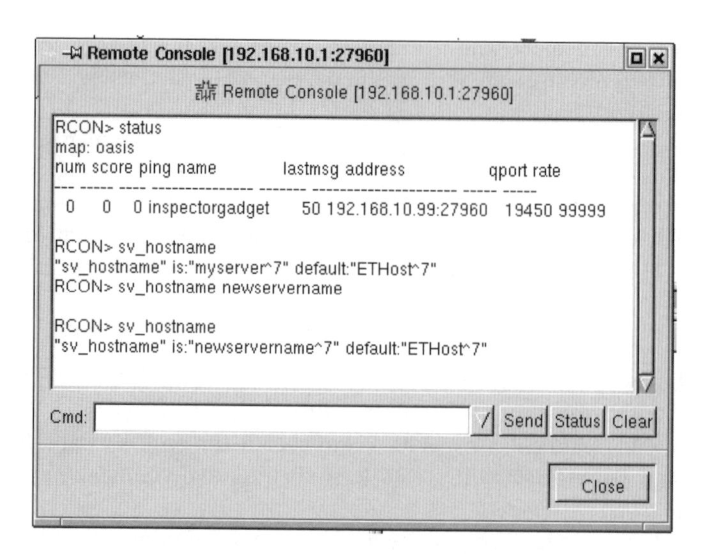

Figure 12.5 Using Rcon within XQF to access a server at 192.168.10.1:27960 (client logged in from 192.168.10.99)

in clear-text over the Internet along with the requested server-side command. (The actual rcon command is sent as an ASCII text string 'rcon ⟨rconpasswd⟩⟨rcon_command⟩' in a UDP packet to the game server's active port.) Anyone sniffing the IP network between your client and your server will easily be able to extract your rcon password and begin remotely interfering with your server. For this reason, rcon should only be used from client machines that have direct IP connectivity to the server (or at least have connectivity over a path that can be trusted).

12.2.8 Automatic Downloading of Maps and Mods

New maps and 'mods' (extensions to the base game engine) are often released by third-party developers. If you run your server with one or more new maps, or a game mod, clients cannot connect unless they too have the same maps and mods installed at their end. Clients can manually retrieve and install the necessary maps and mods before connecting to your server, or the server itself can automatically download the necessary maps and mods when the client first connects. Server downloads of new maps and mods is controlled by the 'sv_allowdownload' option – set it to '1' (the default) to enable downloads, or '0' to block downloads (the client must be manually updated).

Traditional downloading involves data being streamed out to the client by the game server itself, and is usually capped at the snapshot rate (in bytes/second) requested by the client. (And in any case some online discussion of WET suggest the maximum achievable rate is around 13 Kbyte/sec.) Although appropriate for game-play, this can be quite slow for downloading files hundreds or thousands of kilobytes long. (And if you want to further limit the impact of map downloads on the game server, set 'sv_dl_maxRate' to an even lower value.) If you are connected via a LAN or other high-speed link to the server, WET offers an alternative approach – downloads from a separate web site.

Setting 'sv_wwwDownload' to 1 will enable your server to redirect clients to a web site from which they can download new maps and mods. The actual web site is specified by setting 'sv_wwwBaseURL'. For example:

```
+set sv_wwwBaseURL ''http://www.mydomain.com/mymaps''
```

would cause new clients to retrieve additional maps and mods from 'http://www.mydomain.com/mymaps/etmain/' (in other words, the 'mymaps/etmain/' directory on the www.mydomain.com website). Since the webserver can be an entirely different machine to the game server (and indeed may be anywhere else on the Internet), this approach can significantly improve download speeds for clients attaching to your server for the first time.

If the redirection fails, and 'sv_wwwFallbackURL' contains a valid URL, the client will attempt to display the webpage referred to by 'sv_wwwFallbackURL'. This backup webpage should contain instructions on how to manually retrieve the maps and/or mods used by your game server.

12.2.9 Network Performance Configuration

Table 12.2 lists a number of server-side variables and settings affecting the network traffic generated by your server.

Server-side variables can be set on the command line or in server.cfg file.

Table 12.3 lists a number of client-side variables and settings affecting the network traffic generated by your server.

The settings in Table 12.2 on the server side can override settings in Table 12.3 requested at the client side. Chapter 10 provides a discussion of the underlying principles.

12.2.10 Running a Windows Server

In principle, you can start a 'listen server' from within a WET client on a Windows machine. However, this is really only suitable for private, local games. To host a public

Table 12.2 Server-side settings controlling network traffic

Variable	Role	Default
sv_maxrate	Upper bound on 'rate' (the maximum snapshot rate in bytes/second requested by the client)	0 (unlimited)
sv_minping	Clients with ping below this value will not be allowed to join the server	0
sv_maxping	Clients with ping above this value will not be allowed to join the server	0 (unlimited)
sv_lanforcerate	1: If clients are on a local LAN, force their rate and snaps settings to highest possible. 0: disable this feature	1
sv_allowdownload	1: Allow download of new maps and mods from server. 0: disable this feature	1
sv_dl_maxrate	Rate limit in bytes/second for downloading maps and mods to clients (if sv_allowdownload is 1)	42,000 (although in practice often lower)

Table 12.3 Client-side settings controlling network traffic

Variable	Role	Default
Rate	Maximum data rate (in bytes per second) the client wishes to receive snapshots from the server	5000–25,000 (depends on 'network speed')
cl_maxpackets	Rate (in packets per second) at which client will send command packets to the server	30
Snaps	Rate (in packets per second) at which client wishes to receive snapshots from server	20

server on a Windows machine, you should launch the Windows equivalent of the dedicated Linux server described in this chapter. Most of the differences exist only in the name of the executable used to launch the server, and the syntax of filenames and pathnames.

Installation involves downloading and running the version 2.55 installer, and then downloading and running the 2.60 patch (see Table 12.1). Assuming you took the default installation locations, your WET client and server files are most likely stored under 'C:\Program Files\Wolfenstein – Enemy Territory\', and the dedicated server executable is 'ETDED.exe'.

When launched from the command line, ETDED.exe uses the same command line options as previously described for the dedicated Linux server. The hierarchy of folders and sub-folders under C:\Program Files\Wolfenstein – Enemy Territory\' replicates that of the Linux installation.

12.2.11 Further Reading

It is not really within the scope of this book to provide a tutorial on tweaking server configuration options, as most of them relate to WET game-play, in-game physics, and the PunkBuster anti-cheat system [PBUSTER]. Documentation installed along with WET itself (mentioned earlier) provides a starting point for further reading, and there are numerous sites around the Internet with suggestions on how to tweak game-play and add 'mods' (modifications) such as ETPro [ETPRO05].

12.3 Half-Life 2

'Half-Life 2' was released as a single-player FPS game in late 2004 by Valve [VALV2005], an innovative successor to the original Half-Life [HALFLIFE2004]. Included in the client software, 'Half-Life 2: Deathmatch' allows multiplayer, online death-match style of play. An upgraded version of Counter-Strike, the team-play modification of Half-Life, was also released as 'Counter-Strike: Source'.

One of the most significant aspects of Half-Life 2 was Valve's launch of Steam – an entirely revamped client authentication and software distribution system [STEAM2004]. All players must establish Steam accounts, which are then used for authentication and validation purposes (even when playing the single-player mission). A player's legitimately purchased copies of Half-Life 2, Counter-Strike:Source, etc., are registered against the player's Steam account. With their Steam account a player can also purchase additional games online without ever holding a CD or DVD again.

Installing Half-Life 2 on a home PC involves installing a Steam client that subsequently auto-launches whenever the PC restarts. By making Steam an integral part of the Half-Life 2 experience, Valve also ensure that patches to their flagship games are pushed out as uniformly and expeditiously as possible.

In this section, we will discuss the installation and set-up of a dedicated Half-Life 2 multiplayer (HL2MP) server that can be reached by players from around the Internet.

12.3.1 Obtaining and Installing the Linux Dedicated Server

Valve's Steam service actually simplifies the steps you need to know for initially installing the dedicated server. For now, we will assume that your intended server host has a keyboard and monitor and is running at least a basic text console (the installer does not use any menus and does not require a GUI).

Create (or ask the system administrator to create) a dedicated user account for running the HL2MP server (for example, user name 'hl 2' and a home directory of '/home/hl2/'). This helps to keep all your HL2MP server activities distinct from other users on the server host. HL2MP does not need root (administrator) privileges to run. For the rest of this description you are assumed to be logged in as the 'hl 2' user.

Figure 12.6 shows the steps for initially retrieving the Steam client. Use the command 'fetch' (or a similar command 'wget') to retrieve hldsupdatetool.bin from the main Steam site. Make the file executable ('chmod 755'), then execute it. This extracts a new executable called 'steam'. Run this once without any command line parameters, and it will automatically update itself over the Internet. If you were to now re-run './steam' without parameters, you would get its usage message on the screen.

Figure 12.7 shows the command line you must now use to begin installing the HL2MP server itself. These particular options tell the Steam client to 'update' (and download in its entirety if necessary) the hl2mp server components and place them in directory 'hl2mp' (relative to the current directory, i.e. '/home/hl2/hl2mp'). The final line shows you what files and subdirectories have been installed. At the time of writing, the full installation takes around 671 Mbytes of disk space.

You can re-run the Steam client at any time to check for updates. Figure 12.8 shows the output when run immediately after our first installation – there is nothing to update. (Of course, by the time you read this the specific versions of each component are likely to be different.)

```
/home/hl2% fetch http://www.steampowered.com/download/hldsupdatetool.bin
/home/hl2% chmod 755 ./hldsupdatetool.bin
/home/hl2% ./hldsupdatetool.bin
/home/hl2% ./steam
Checking bootstrapper version ...
Getting version 14 of Steam HLDS Update Tool
Downloading. . . . . . . . . . . .
Steam Linux Client updated, please retry the command
/home/hl2%
```

Figure 12.6 Using the Steam installer to update itself

```
/home/hl2% ./steam -command update -game hl2mp -dir hl2mp
    [..steam downloads roughly 671MByte of hl2 and hl2mp material..]
/home/hl2% ls -al hl2mp
total 562
drwxr-xr-x   5 hl2  hl2      512 Oct 24 21:50 .
drwxr-xr-x   5 hl2  hl2      512 Oct 24 21:50 ..
-rw-r--r--   1 hl2  hl2     1381 Oct 24 21:52 InstallRecord.blob
drwxr-x---   2 hl2  hl2     1024 Oct 23 23:36 bin
drwxr-x---   7 hl2  hl2      512 Oct 23 23:36 hl2
drwxr-x---  11 hl2  hl2      512 Oct 23 23:24 hl2mp
-rwxr-xr--   1 hl2  hl2   183825 Oct 23 23:36 srcds_amd
-rwxr-xr--   1 hl2  hl2   183793 Oct 23 23:36 srcds_i486
-rwxr-xr--   1 hl2  hl2   183793 Oct 23 23:36 srcds_i686
-rwxr-xr--   1 hl2  hl2    10164 Oct 23 23:36 srcds_run
/home/hl2%
```

Figure 12.7 Using the Steam installer to install a Half-Life 2 multiplayer server

```
/home/hl2% ./steam -command update -game hl2mp -dir hl2mp
Checking bootstrapper version ...
Updating Installation
Checking/Installing 'Half-Life 2 Deathmatch' version 11
Checking/Installing 'Base Source Shared Materials' version 7
Checking/Installing 'Base Source Shared Models' version 3
Checking/Installing 'Base Source Shared Sounds' version 3
Checking/Installing 'Source Dedicated Server Linux' version 47
HLDS installation up to date
/home/hl2%
```

Figure 12.8 Updating a current hl2mp installation

If you want to install Counter-Strike:Source instead (or in addition to), then replace '-game hl2mp' with '-game "Counter-Strike Source"' in the examples above. You must also specify an alternative directory location with the '-dir' option (otherwise the Counter-Strike:Source software components will likely overwrite some of your hl2mp software components).

Because the Steam installer uses a text-based interface you may also do this installation process while remotely logged into the server host from elsewhere (for example, over an ssh connection [SSH2005]). This is most convenient when your server host is already mounted in remote rackspace without a dedicated keyboard and monitor.

12.3.2 Starting the Server for Public Use

A couple of technical decisions need to be made before starting your dedicated server.

- What UDP port will it run on?
- Is this for local LAN play, or available for anyone on the Internet?
- What is the maximum number of supportable players?

(There are also some game-specific questions you need to consider, such as 'what game type and map rotation do I want?'. In part, these are subjective decisions that depend largely on your motives for running a server. Detailed descriptions of how to set up various game types are outside the scope of this book.)

By default, an HL2MP server will bind itself to UDP port 27015 on the local server host. However, you can explicitly specify an alternative port if another process on the server host is already using port 27015 (for example, another HL2MP server).

Player slots should be limited based on your available network capacity and processor speed. Use the discussion in Chapter 10 to estimate your per-player network bandwidth requirements.

From within the hl2mp subdirectory, the usual way to start a dedicated server is to execute 'srcds_run' with a number of command line options. For example:

```
/home/hl2/hl2mp/% ./srcds_run -game hl2mp -console +map
   dm_overwatch
```

This particular set of options instructs the server to run the hl2mp game on the default UDP port 27015 and start map 'dm_overwatch'. (If no map is specified the server initialises, but does not start operation.) Because of how srcds_run is written you must run this command from within the directory in which you installed the HL2MP server.

By default, the server will look for a configuration file named 'server.cfg' under './hl2mp/cfg'(relative to the directory in which srcds_run is located, which in our example would be at '/home/hl2/hl2mp/hl2mp/cfg/server.cfg'). Almost everything that can be configured about the HL2MP server's operation can be specified or overridden in server.cfg.

By default, the server will start with two player slots. Adding the '+maxplayers N' option tells the server to create N player slots.

If you are running on a host with multiple IP interfaces, you may wish to explicitly identify which IP interface should be advertised by the HL2MP server. For example, the following command line specifies that the server advertises itself (and only 'exists' on) IP address 192.168.1.50:

```
/home/hl2/hl2mp/% ./srcds_run -game hl2mp -console +map
   dm_overwatch -ip 192.168.1.50
```

To run your server on a non-standard UDP port, and bind to a particular IP interface on your local server host, use both the '-ip' and '-port' options. For example, to use port 27085 as the port for in-game traffic in a game for 16 players:

```
/home/hl2/hl2mp/% ./srcds_run -game hl2mp -console +map
   dm_overwatch -ip 192.168.1.50 -port 27085 +maxplayers 16
```

By default, your server will register with Steam and become available to other players around the Internet. If you happen to be sniffing the network connected to your server, you will see server-discovery probe traffic coming in within minutes (or sometimes only seconds) of starting your server. (You may also see server-discovery probes from the Internet for some period of time after you shut down your public server. It takes time for knowledge of your server's IP and port number to disappear from all the places it was stored when you first registered as a public server.)

If you copy the './steam' executable (used to install and update HL2MP) into the same directory as 'srcds_run', you can use the '-autoupdate' command line option to request that srcds_run updates the HL2MP server immediately prior to starting. (Internet connectivity is required for autoupdate to work. This option will not be very helpful if you are running a LAN-only server on an isolated network.)

12.3.3 Starting a LAN-only Server

Adding the '+sv_lan 1' option prevents your server from registering as a public server. For example:

```
/home/hl2/hl2mp/% ./srcds_run -game hl2mp -console +map
   dm_overwatch +sv_lan 1
```

Your server will still contact the Steam master servers, but will not register as a public server, and will not use Steam to authenticate clients and will not use Valve Anti-Cheat (VAC) to protect the server. The server may be located by players on your local LAN using the 'LAN game' option in their Steam clients.

Clients discover servers on their local LAN by broadcasting server-discovery probes (UDP packets sent to IP address 255.255.255.255 with the ASCII text 'TSource Engine Query' in their payloads). Each time a player presses their 'refresh' button in the Steam client's LAN server locator window, six UDP broadcast probes are sent – one each to ports 27015, 27016, 27017, 27018, 27019 and 27020 in that order. This normally triggers a response from any of the dedicated servers on the local LAN who are listening on one of these six common HL2MP server ports. A server using, for example, port 27085 would *not* be visible to the Steam client's LAN search function. A prospective player would need to explicitly name the server by IP address and port number in order to connect.

As with WET (discussed earlier in this chapter) if you run a LAN-only game using '+sv_lan 1′ and also specify the game server's local IP address with '-ip w.x.y.z' the server appears to ignore the server-discovery queries being sent to 255.255.255.255. In this case, each client will need to manually specify the server's IP address (w.x.y.z) and port number in order to connect.

12.3.4 Ports You Need Open on Firewalls

It is entirely possible you have a firewall between your server and the outside world. If so, you must ensure that the server can send packets outbound to a range of ports. HL2MP seems to need quite a few ports open for clients and servers to operate freely. A dedicated public server has been seen to talk to remote Steam-related servers on ports 27011 (UDP), 27014 (UDP) and 27030 (TCP). UDP port 27020 is required open for Half-Life TV. Inbound access is obviously required to the server's actual game-play port (UDP to 27015, or whatever was set with '-port NNN'). If remote administrative access is required (using the 'rcon' function, discussed later) you will also need to enable inbound TCP access to the game-play port.

12.3.5 Dealing with Network Address Translation

Running an HL2MP server behind a NAT-enabled router introduces a number of issues. These have largely been discussed earlier in this chapter in relation to WET, and will not be repeated here. Although HL2MP and Steam use a number of ports, only the gameplay port ('-port NNNN') must be visible to the outside world. For the reasons described earlier, you will need to install a special rule into your NAT-enabled router to ensure unsolicited server-discovery probes from the wider Internet are properly routed to your HL2MP server. If you plan to remotely manage your server with 'rcon', you will need to put in the same rule for both UDP and TCP traffic.

12.3.6 Monitoring and Administration

When started from the command line the game server prints information to the screen and accepts keyboard input in real time, allowing you to monitor server state and modify server configuration parameters. For example, you could type 'status' to find out the names of all clients currently connected, along with their current ping time, IP address and port number.

However, you'll usually want to start the game server and then leave – returning only occasionally to monitor activities and perhaps modify parameters. For this you need to enable logging and 'rcon' (remote console).

To set up logging from your server, the following commands should be in your server.cfg file:

```
log on
sv_logsdir "<logdirectoryname>"
sv_logfile 1
sv_logecho 1
sv_logdetail 3
sv_logmessages 1
```

The ⟨logdirectoryname⟩ parameter is the name of a directory within which you wish to store your logfiles. If provided as a relative pathname, it is interpreted relative to the hl2mp game directory. For example, when using the hl2 user account described above, 'sv_logsdir "logs/todayslogs"' would result in logfiles being created inside the directory '/home/hl2/hl2mp/logs/todayslogs/'.

The actual logfiles will be created inside this directory with names 'Lmmddxxx.log', where mm is the month, dd is the day, and xxx in an integer number starting at '000' and incrementing with every map change. Each logfile will contain date and time stamps of events such as players joining and leaving, their playernames, Steam IDs and the IP address:port from which the client connected.

Remote control of your game server is possible through the 'rcon' facility. Rcon is a client-side console command that passes text to the server, and prints the server's output back onto the client's console window. Rcon access to a server is enabled by setting an rcon password with the '+rcon_password ⟨password⟩' option when starting your server.

For example, to start a LAN-only server with the rcon password of 'mypass' you could use:

```
/home/hl2/hl2mp/% ./srcds_run -game hl2mp -console +map
    dm_overwatch +sv_lan 1 +rcon_password mypass
```

Figure 12.9 shows how you would then access the server from the console of an HL2 game client already connected to the server. The first use of 'rcon status' resulted in

```
Console
] rcon status
Bad RCON password
] rcon_password mypass
] rcon status
hostname:  Half-Life 2 Deathmatch
version   : 1.0.0.7/7 2536 insecure
udp/ip   :  192.168.10.1:27015
map      :  dm_overwatch at: 0 x, 0 y, 0 z
players :  1 (2 max)

# userid name uniqueid connected ping loss state adr
#  2 "grenville" STEAM_ID_LAN 04:15 44 0 active 192.168.10.99:27005
] rcon sv_lan
"sv_lan" = "1" ( def. "0" )
 - Server is a lan server ( no heartbeat, no authentication, no non-class C addresses )
]
] screenshot
Submit
```

Figure 12.9 Using Rcon from within Half-Life 2 client console to access a server at 192.168.10.1: 27015 (client logged in from 192.168.10.99)

an error message because the server requires an rcon password. So you set the rcon password with 'rcon_password mypass', and then repeat the 'rcon status' command. Now you receive the server's console output back (seeing that there's one person, the author, on this LAN-only server). You can issue other commands too, such as 'rcon sv_lan' to report the status of the LAN-only flag.

You only need to enter the 'rconpassword' line once, then use multiple 'rcon' commands as necessary.

HL2's rcon service runs over TCP to the server's game-play port (27015, unless '-port' was specified when the server started up). Unfortunately, the rcon mechanism has an important weakness. Every time an rcon command is executed, the client sends the rcon-password in clear-text over the Internet along with the requested server-side command. Anyone sniffing the IP network between your client and your server will easily be able to extract your rcon password and begin remotely interfering with your server. For this reason, rcon should only be used from client machines that have direct IP connectivity to the server (or at least have connectivity over a path that can be trusted).

At the time of writing XQF [XQF05] did not support 'rcon' to HL2MP servers, although this may well have been added subsequently.

12.3.7 Network Performance Configuration

Table 12.4 lists a number of server-side variables and settings affecting the network traffic generated by your server.

Server-side variables can be set on the command line or in server.cfg file (except for '-tickrate', which must be set on the command line).

Table 12.5 lists a number of client-side variables and settings affecting the network traffic generated by your server.

The settings in Table 12.4 on the server side can override settings in Table 12.5 requested at the client side. Chapter 10 provides a discussion of the underlying principles.

12.3.8 Running a Windows Server

In principle, you can start a 'listen server' from within a Half-Life 2 client on a Windows machine. However, this is really only suitable for private, local games. To host a public

Table 12.4 Server-side settings controlling network traffic

Variable	Role	Default
sv_minrate	Lower bound on 'rate' (the maximum snapshot rate in bytes/second requested by the client)	0
sv_maxrate	Upper bound on 'rate' (the maximum snapshot rate in bytes/second requested by the client)	0 (unlimited)
sv_minupdaterate	Lower bound on cl_updaterate (the client-side requested snapshot rate). Server will send at this rate if cl_updaterate is too low	10
sv_maxupdaterate	Upper bound on cl_updaterate (the client-side requested snapshot rate)	60
-tickrate	Sets the server's internal tick rate (in ticks per second). Can only be set on the command line	66

Table 12.5 Client-side settings controlling network traffic

Variable	Role	Default
rate	Maximum data rate (in bytes per second) the client wishes to receive snapshot data from the server	(depends on client 'network' setting)
cl_cmdrate	Rate (in packets per second) at which client will send command packets to the server	30
cl_updaterate	Rate (in packets per second) at which client wishes to receive snapshots from server	20

server on a Windows machine, you should download the Windows equivalent of the dedicated Linux server described in this chapter. Most of the differences exist only in the name of the executable used to launch the server, and the syntax of filenames and pathnames.

Valve provides an automatic update tool called *hldsupdatetool.exe*, located at http:// www. steampowered.com/download/hldsupdatetool.exe. Download this tool and launch it without any options. It will update itself, and is then ready to be used to pull down the server itself. You will be prompted by hldsupdatetool.exe to specify a directory into which you would like to store your dedicated server. Two obvious choices would be C:\srcds\ or C:\hl2mp\ – it is up to you. Once the tool has updated itself, launch it again with the same options as for the Linux updater, e.g.:

```
hldsupdatetool.exe -command update -game hl2mp -dir c:\hl2server
```

(The primary difference is how we locate and specify directories.)

Once all files have been downloaded, you can launch the dedicated server in a similar fashion to the Linux server. For example, to launch HL2MP with map dm_overwatch, LAN-only and rcon password of 'mypass', use this line:

```
c:\hl2server\srcds.exe -game hl2mp -console +map dm_overwatch
   +sv_lan 1 +rcon_password mypass
```

You may find it useful to create a short batch file that loops forever, automatically restarting the server if it crashes for some reason.

12.3.9 Further Reading

It is not really within the scope of this book to provide a tutorial on tweaking server configuration options, as most of them relate to HL2MP game-play, in-game physics, and the VAC system. There are numerous sites around the Internet with suggestions on how to tweak game-play and add 'mods' (modifications). A good place to start would be Valve's own Steam Forums at http://steampowered.com.

12.4 Configuring FreeBSD's Linux-compatibility Mode

FreeBSD is popular in many server configurations. With its Linux-compatibility mode FreeBSD makes a good platform for hosting dedicated game servers. In this section, we will discuss two key steps that must be performed before running a dedicated Linux game server under FreeBSD. The first is to ensure the correct Linux-compatibility libraries are installed. The second is to ensure the FreeBSD kernel is 'ticking' fast enough.

12.4.1 Installing the Correct Linux-compatibility Libraries

Details of FreeBSD's Linux binary compatibility are described in Chapter 10 of the online FreeBSD documentation [FBLINUX] and will not be repeated here. FreeBSD 5.4 was current at the time of writing, and a number of Linux-compatibility options exist.

Linux binary compatibility means that a Linux application (an 'executable' binary file previously compiled under Linux) runs natively on a FreeBSD host. There is no processor emulation occurring – compatibility is implemented within the FreeBSD kernel's loader function. When a Linux application is launched, the FreeBSD kernel will re-write sections of the application's internal executable code before starting the program. (The Linux application's calls into the Linux kernel for things like disk access, network access and memory management are replaced with equivalent calls into the FreeBSD kernel.) After the re-writes are complete, the application begins execution as though nothing has happened. Many standard Linux libraries are installed under '/compat/linux' on the FreeBSD host, enabling dynamically linked applications to operate properly.

The main point to recognise is that Linux compatibility is easily configured by installing a 'Port' or 'Package' onto your existing FreeBSD system. A number of different compatibility environments are available, reflecting some different Linux distributions. If you want to run Half-Life 2 dedicated server you will need the libraries based on Red Hat Linux 8.0, rather than the default (as of FreeBSD 5.4) based on Red Hat Linux 7.x.

If you are already connected to the Internet and logged in as root, simply execute:

```
% pkg_add -r linux_base-8
```

All the necessary files will be automatically retrieved over the Internet to install a compatibility environment based on Red Hat Linux 8.0. This package will also support the WET dedicated Linux server.

So, if you have a FreeBSD server available it is entirely reasonable to host Linux-based dedicated game servers on this machine.

12.4.2 Ensuring the Kernel 'Ticks' Fast Enough

Most operating systems have some form of internal 'software clock' ticking along at a fixed rate, used by programs that wish to regularly 'wake up and do something'. This is true of Windows, Linux and FreeBSD. (This tick rate is not to be confused with a game server's own internal tick rate, discussed earlier in the book. The two tick rates can be entirely different.)

At least up until FreeBSD 5.4, the kernel's own tick timer defaulted to 100 ticks per second. Given the way dedicated game servers are written, this tick rate is way too low.

Fortunately, the FreeBSD tick rate can be adjusted without recompiling the kernel. Simply add the following line to the file /boot/loader.conf and reboot the machine.

```
kern.hz="1000"
```

Once rebooted the kernel will be ticking 1000 times a second, a much better rate when hosting game servers. (You could even set it faster – the limit really depends on how much CPU load your machine can tolerate. There is a small, fixed amount of processing overhead incurred for each tick of the kernel's clock, regardless of what all the applications are doing.)

You can check or confirm your kernel's current tick rate with the following command:

```
sysctl kern.clockrate
```

Increasing the kernel's own tick rate is important for any system hosting an online dedicated game server, whether it be Linux, FreeBSD or Windows. The simple fact is that a game server's own activities are generally limited by the kernel's own tick timer.

For example, an HL2MP dedicated server on a standard FreeBSD 5.4 host is limited to sending at most (kern.clockrate/2) snapshots per second to each client, regardless of the game server's own tickrate (which defaults to 66). Given a default kernel tick rate of 100, your HL2MP server would be unable to send more than 50 snapshots per second to each client. The game server's own default sv_maxupdaterate setting of 60, and default tick rate of 66, are rendered meaningless.

Running a FreeBSD server host at, for example, 150 ticks per second would limit the HL2MP server to 75 snapshots per second to each client. Setting the kernel's tick rate to 1000 ensures that sv_maxupdaterate, cl_updaterate and the HL2MP game server's own tickrate become the proper limiting factors.

In summary: the maximum snapshots per second rate to each client is the smaller of sv_maxupdaterate, the game server's tickrate, and half the underlying operating system's software tick rate. It is essential to reconfigure your FreeBSD's default 100 ticks per second to something more like 1000 ticks per second.

References

[DUMMYNET] dummynet, “Dummynet – Traffic Shaper, Bandwidth Manager and Delay Emulator”, http://www.FreeBSD.org/cgi/man.cgi?query=dummynet&sektion=4, 2004.

[ETPRO05] ET Pro, “ET Pro – The Enemy Territory Competition Mod”, http://etpro.anime.net/.

[FBLINUX] FreeBSD Handbook, “Chapter 10 Linux Binary Compatibility”, (http://www.freebsd.org/doc/en_US.ISO8859-1/books/handbook), 2005.

[HALFLIFE2004] H A L F – L I F E 2, http://half-life2.com, 2004.

[LEE2005] S. Zander, I. Leeder and G. Armitage, “Achieving Fairness in Multiplayer Network Games through Automated Latency Balancing”, ACM SIGCHI International Conference on Advances in Computer Entertainment Technology (ACE2005), Valencia,Spain, June 2005.

[PBUSTER] PunkBuster, “PunkBuster Online Countermeasures,” http://www.punkbuster.com/, Accessed 2006.

[RIZ1997] L. Rizzo, “Dummynet: A simple approach to the evaluation of network protocols”, *ACM Computer Communication Review*, Vol. **27**, No. 1, pp. 31–41, 1997.

[SSH2005] OpenSSH, http://www.openssh.org, 2005.

[STEAM2004] Welcome to Steam, http://www.steampowered.com/, 2004.

[VALV2005] Valve, http://www.valvesoftware.com, 2005.

[WET2005] Wolfenstein, http://games.activision.com/games/wolfenstein as of October 2005.

[XQF05] XQF Game Server Browser, http://www.linuxgames.com/xqf/index.shtml, 2005.

13

Conclusion

By the end of a book both the authors and readers realise there is much more 'out there' that we have not, and simply could not, cover. We can be sure of one thing in this field of networking and online games – the specific technologies and games will continue to evolve rapidly and have a deeper impact on society, but the underlying issues and principles will remain. To that end we conclude with pointers to a small collection of online resources – places to find the latest news, more technical details, and forums discussing game and networking technologies.

13.1 Networking Fundamentals

We have provided an introduction to TCP, UDP and IP networking in this book, without stressing too much on the details. A number of books are available that focus specifically on IP networking issues, including the following well-known texts:

- "Internetworking with TCP/IP Vol. I: Principles, Protocols, and Architecture" by Douglas E. Comer (5th edition June 30, 2005) Prentice Hall; ISBN: 0131876716
- "TCP/IP Illustrated, Volume 1: The Protocols" by W. Richard Stevens (January 1994) Addison-Wesley Pub Co; ISBN: 0201633469
- "Computer Networks" by Andrew S. Tanenbaum (4th edition August 9, 2002) Prentice Hall; ISBN: 0130661023

Various sources of online information exist. The Internet Engineering Task Force (IETF) and Institute of Electrical and Electronic Engineers (IEEE) both have websites where you can obtain copies of their standards documents describing a wide variety of Internet and link layer technologies.

- Internet Engineering Task Force (IETF)
  - http://www.ietf.org
- Institute of Electrical and Electronic Engineers (IEEE)
  - http://www.ieee.org
- IEEE Working Group for Wireless LANs
  - http://www.ieee802.org/11

IETF documents crop up in a number of chapters, known as *'RFCs'* (from their traditional name, 'Requests for Comment'). A somewhat user-unfriendly search tool is available at http://www.ietf.org/rfc.html (you must know the RFC number), or go to http://www.rfc-editor.org/to search on various text fields.

The Association for Computing Machinery (ACM, http://www.acm.org) also provides a wide range of publications covering networking and computing topics. A number of ACM publications and special interest group (SIG) journals and conference proceedings touch on issues relevant to networked game development. These include:

- SIGCHI (http://www.acm.org/sigchi)
- Computers In Entertainment (http://www.acm.org/pubs/cie.html)
- SIGCOMM (http://www.acm.org/sigcomm)
- SIGGRAPH (http://www.siggraph.org/)

13.2 Game Technologies and Development

A number of online websites and resources exist, covering various aspects of game distribution, research, development and design. These include the following:

- http://www.theesa.com/
  - Entertainment Software Association: 'The ESA works with the government at all levels to make the voice of its members heard on a wide range of crucial legislative and public policy issues, including intellectual property protection, content regulation, and efforts to regulate the Internet'.
- http://www.gamasutra.com/
  - Gamasutra: 'The Art & Business of making games'. Currently an excellent and active site for all things to do with game development, the games industry and game technology.
- http://www.gamespy.com/
  - GameSpy: hosts a wide variety of game-specific forums, file download areas, and so on (consoles and PCs).
- http://www.igda.org/
  - 'The International Game Developers Association is a non-profit professional membership organization that advocates globally on issues related to digital game creation'.
- http://www.gamedev.net/
  - GameDev.net: 'all your game development needs'. Another site with information for game developers.
- http://www.gdconf.com/
  - Game Developer's Conference
- http://www.gamespot.com/
  - Gamespot.
- http://www.3dgamers.com/
  - 3D Gamers (news, downloads)
- http://www.digiplay.org.uk/
  - Digiplay Initiative: 'Research into computer gamers and the industry they are part of'.

13.3 A Note Regarding Online Sources

It has become commonplace to utilise online resources to find answers to questions, example code or opinions on all sorts of topics. Search engines, such as Google (www.google.com) and Yahoo (www.yahoo.com), have become an entry point to a vast range of websites with information. Community-edited resources, such as Wikipedia ('The Free Online Encyclopedia', www.wikipedia.com), have become fascinating sources of facts and fiction. Information can be of high quality (created by people who care about the accuracy and relevance of their material) or of entirely dubious quality (created by people concerned more with opinion than accuracy, or created months and years in the past and subsequently never updated to reflect a changing reality).

The best sources of information are normative – where the website is run by the organisation who created the facts or opinions you wish to read and/or cite. Secondary sources, those who summarise or purport to reflect the facts of a situation they did not create or own, can be reliable to a degree. You should always consider carefully the nature of your source's relationship to the information and facts about which your source purports to report. Where normative sources exist, they should always be utilised in preference to secondary sources. Search engines can be very helpful in locating normative sources for information you may have initially discovered through a secondary source (such as an online newspaper or game developer's forum).

Keep in mind that web-based content can change in seconds, so the information you saw at a particular URL at time X may have changed when someone else accesses the site sometime later. For example, Wikipedia cautions readers to keep track of precisely when they saw a particular version of any given article, because they can change so frequently (http://en.wikipedia.org/wiki/Wikipedia:Citing_Wikipedia). When following an online reference given by someone else, remember that the version you see now may differ from what your secondary source saw in the past.

Index